CREATE

EDUCATIONAL

ROBOTICS

Dr Kesorn Pechrach Weaver

CREATE EDUCATIONAL ROBOTICS

Dr Kesorn Pechrach Weaver

Pechrach Publishing
United Kingdom

Create Educational Robotics

By Dr Kesorn Pechrach Weaver

ISBN 978-1-912957-04-0

PECHRACH PUBLISHING

7 Boundary Road, Bishops Stortford, Hertfordshire, CM23 5LE, England, United Kingdom. Tel: (+44) 1279 508020, +44 (0) 7443426937

Published 2019 by Pechrach Publishing

This book is dedicated to my robotics students

Acknowledgments

I would like to thank to my students from the Computer Science and Robotics Class; Thomas Wilkinson, Ruby Luck, Joshua Williams, Jorja Cox, Harry McCrae, Eleanor Chichlowska, Benjamin de Vos, Aaron Walker, Alex Rodriguez, Deborah Toms, Ellie Simeonova, Felix Baker, Hattie More, Naomi Williams, Ollie Hall, Paige Tritton, Tom Gibbs, Ander Miera, Ben Cooper, Florence Chichlowska, Hannah Walker, Freddy Brazier, Jack Cooper and Scarlett Murray. They are brilliant and provide me with a new vision and dimension of thinking.

I would like to thank M Emlyn Humphries, Lisa Nichols, Rachel Kelly, Kitisak Nullak, Dr. Sirilaksana Kunjara, J.P. Darby and Rung Ratpinyyotip, for all their support.

Thanks Dr. Manop Sittidech, Dr. Choomjet Karnjanakesorn, Dr. Somvong Tragoonrung, Dr. Vanida Khumnirdpetch, Assoc. Prof. Seensiri Sangajit, Asst. Prof. Dr. Niwat Moonpa, Assoc. Prof. Dr. Teerasak Urajaranon, Ms. Sureeporn Yai sa-nga, Dr. Suraphon Chaiwongsar, Asso. Prof. Dr.Somchart Hanvangsa, Dr. Passawat Wacharadumrongsak, Dr.-Ing. Fongjan Jirasit, Dr. Punnee Sittidech, Dr. Kom Silapajarn, Dr. Cherdsak Virapat, Dr. Pongpat Boonchuwong, Dr. Julathep Kajornchaiyakul, Dr. Chai Wutiwiwatchai, Jaturong Amonchaisup, Thanakorn Chanmalee, Puripun Sophastienphong, Benvenido Opiniano, Prof. J.W. McBride, Siriluk Pumirat, Juliette Karen Weaver, Chiraphat Kumpidet, and Assist. Prof. Suphannika Phutthachalee, for giving me an inspiration.

A special thanks to Dr Paul M. Weaver and Neran J.P. Weaver for their support both mentality and physicality during preparing this book.

Many thanks to Dr. Yeetoh Dabbhadatta, Dr. Pantip Ampornrat, Dr. Wirulda Pootakham, Dr. kanokwan Natikool, Dr. Kanjana Tuantet, Prof. Dr. Apichart Vanavichit, Dr. Pattharaporn Suntharasaj, Dr. Phawika Rueannoi, Dr. Sorraya Kagittanon, Dr. Suttirat Rattanapan, Dr. Thitaphat Ngernsutivorakul, Dr. Wilaiporn Chetanachan, Dr. Seetala Jamrerkjang, Dr. Nataporn Chanvarasuth, Dr. Tassanai Sanponpute, Dr. Jackapon Sunthornvarabhas, Dr. Sirion Umarin, Dr. Wirulda Nik Pootakham, Tanat Tonguthaisri, Wipaporn Ekamornthannakul, Sronkanok Tangjaijit,Tipawan Tangjitpiboon,Tharnpraporn Chairatana and Nualanong Tongdee.

Finally, I would like to thank my family in Thailand and my friends and family in the United Kingdom for believing in me.

Table of Contents

Acknowledgement

Table of Figures

CHAPTER 1

Introduction to Arduino

Install Arduino

The first thing we need to do is to download and install the Arduino software into our computer. Go to the web page https://www.arduino.cc/

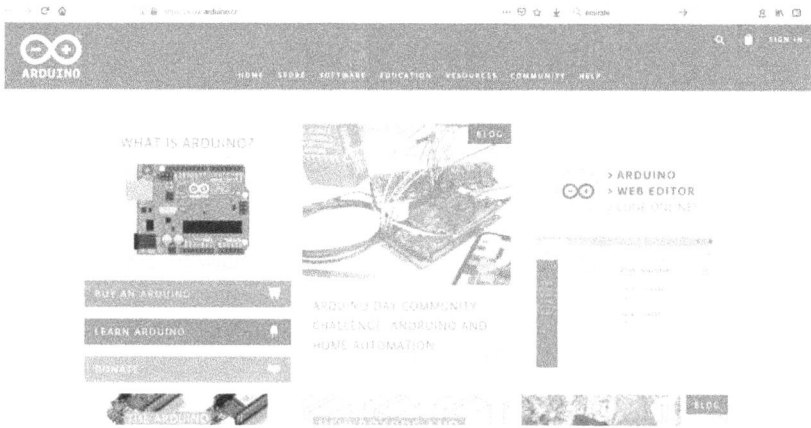

Figure 1.1: Arduino webpage

At the top bar, choose the software and click on the Download link.

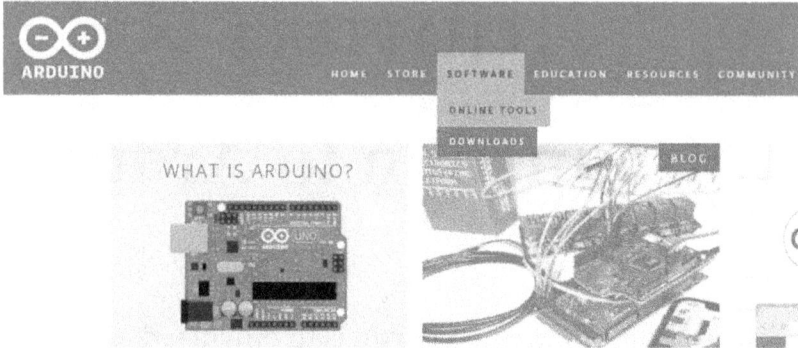

Figure 1.2: Arduino download

There are many versions of Arduino such as for Windows, Linux and Mac.

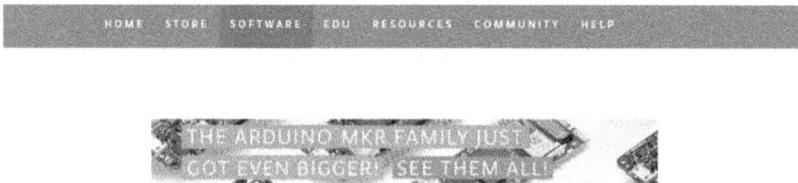

Figure 1.3: Arduino version

There are options to contribute to the Arduino software or just Download.

Figure 1.3: Contribution to the Arduino

When we click " Just Download", the pop up screen would show and ask if we would like to save this file.

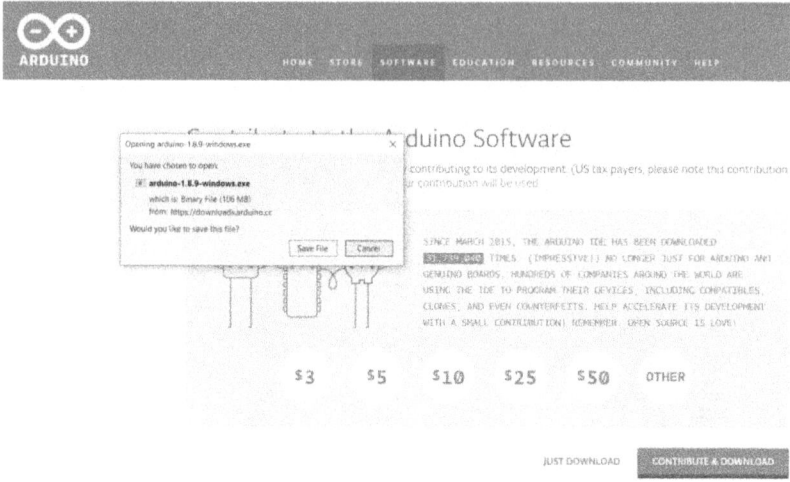

Figure 1.4: Save Arduino file

After we click "Save", the file would be downloaded and in the Download folder.

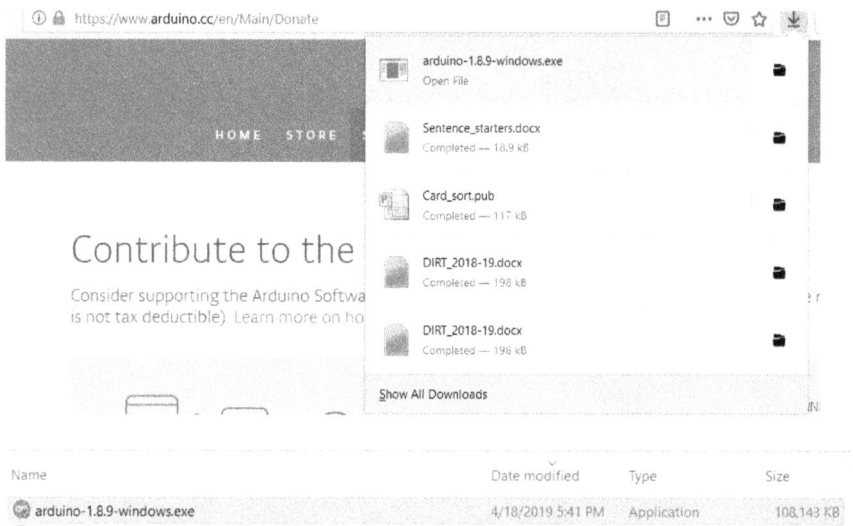

Figure 1.5: Download file

Create Educational Robotics

When double click on the link, the file would start to install and run the Arduino programme as below.

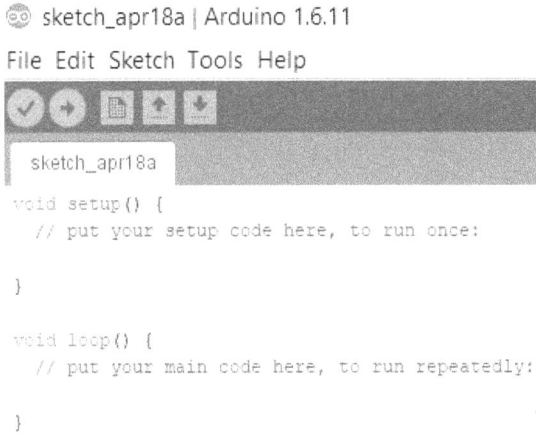

sketch_apr18a | Arduino 1.6.11

File Edit Sketch Tools Help

sketch_apr18a

```
void setup() {
  // put your setup code here, to run once:

}

void loop() {
  // put your main code here, to run repeatedly:

}
```

Figure 1.6: Arduino Program

sketch_apr18a | Arduino 1.6.11

File Edit Sketch Tools Help

New	Ctrl+N
Open...	Ctrl+O
Open Recent	>
Sketchbook	>
Examples	>
Close	Ctrl+W
Save	Ctrl+S
Save As...	Ctrl+Shift+S
Page Setup	Ctrl+Shift+P
Print	Ctrl+P
Preferences	Ctrl+Comma
Quit	Ctrl+Q

Built-in Examples

01.Basics	>	AnalogReadSerial
02.Digital	>	BareMinimum
03.Analog	>	Blink
04.Communication	>	DigitalReadSerial
05.Control	>	Fade
06.Sensors	>	ReadAnalogVoltage
07.Display	>
08.Strings	>
09.USB	>
10.StarterKit_BasicKit	>
11.ArduinoISP	>

Examples from Libraries

Bridge	>
EEPROM	>
Firmata	>
SoftwareSerial	>
SPI	>
Temboo	>
Wire	>
RETIRED	>

Examples from Custom Libraries

Ethernet	>
GSM	>
LiquidCrystal	>
SD	>
Servo	>
Stepper	>
TFT	>
WiFi	>

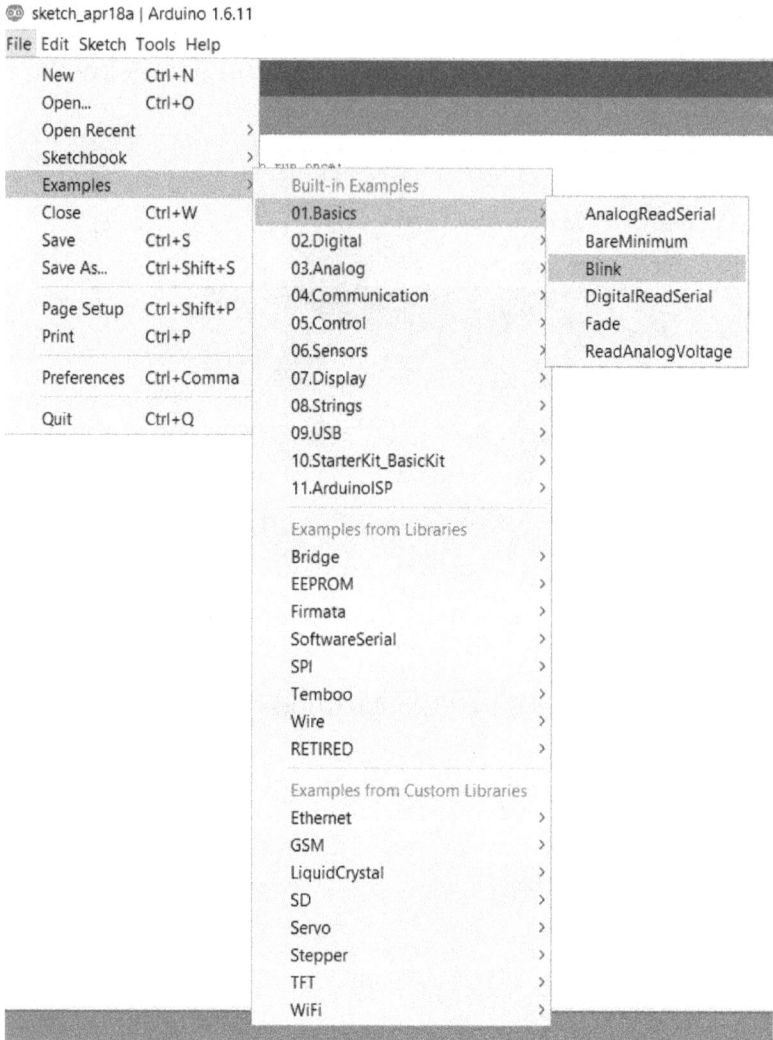

Figure 1.7: Examples Arduino Code

Create Educational Robotics

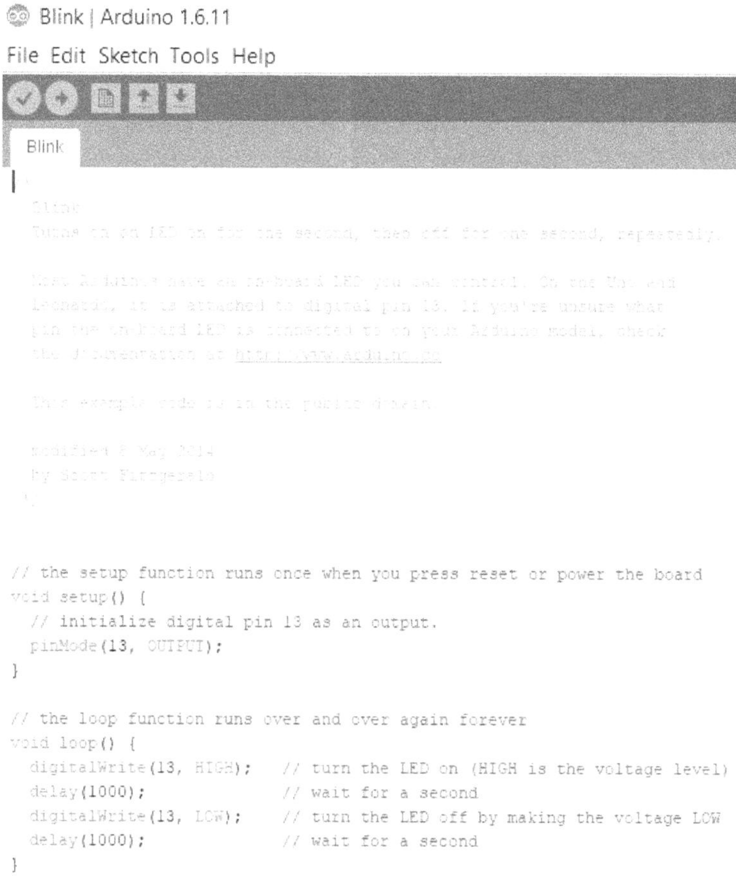

```
Blink | Arduino 1.6.11
File Edit Sketch Tools Help

Blink

// the setup function runs once when you press reset or power the board
void setup() {
  // initialize digital pin 13 as an output.
  pinMode(13, OUTPUT);
}

// the loop function runs over and over again forever
void loop() {
  digitalWrite(13, HIGH);   // turn the LED on (HIGH is the voltage level)
  delay(1000);              // wait for a second
  digitalWrite(13, LOW);    // turn the LED off by making the voltage LOW
  delay(1000);              // wait for a second
}
```

Figure 1.8: Blink Program

There are many example programs trying it on the Arduino board. This includes display, communication and servo motors.

In this book, we would use the Arduino to control motors in the robotics four wheels, three wheels and robotics five toes foot.

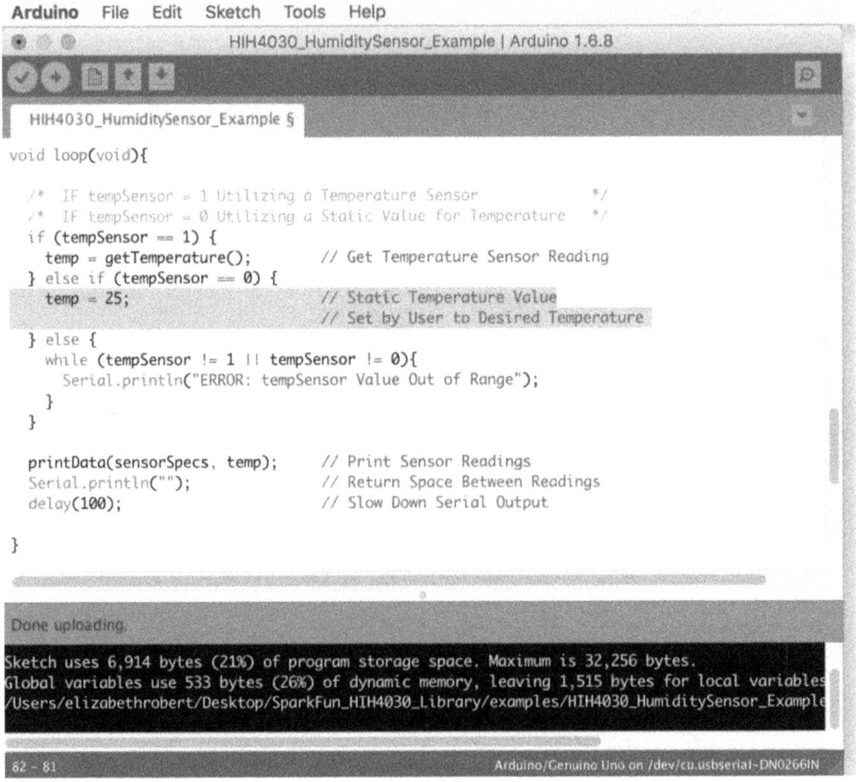

Figure 1.9: Humidity sensor

CHAPTER 2

Equipment

The basic tools, equipment and materials need to prepare before starting building the car robots are shown as follows:

1. Car Chassis

2. Motors

3. Arduino UNO board

4. L298 Motor Drive

5. HC-05 Bluetooth Module

6. Batteries

7. Power Adapter

8. USB cable

9. Jumper wires

10. Crocodile clips

11. Multimeter

12. Screwdriver

13. Mobile phone

14. Solder Iron

15. Solder wire

16. Front Wheel

17. Big Back Wheels

18. Small Wheels

19. Motors Fixtures

20. Front Wheel Fixtures

21. Switch

22. Battery Holder

23. Motor Lock Wheel

24. Two way screw driver

25. Scissor

26. Mini Needle Nose Plier

27. Insulation tape

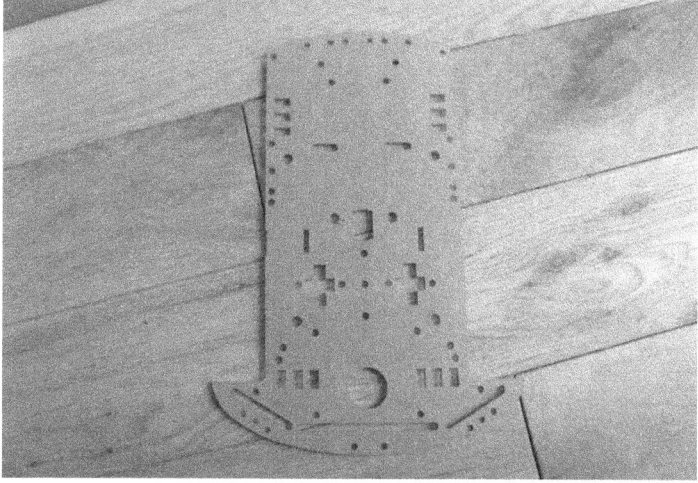

Figure 2.1: Car Chassis

Figure 2.2: Motors

Figure 2.3: Arduino board

Figure 2.4: L298 Motor Drive

Figure 2.5: HC-05 Bluetooth Module

Figure 2.6: Batteries

Figure 2.7: Power Adapter

Figure 2.8: USB Cable

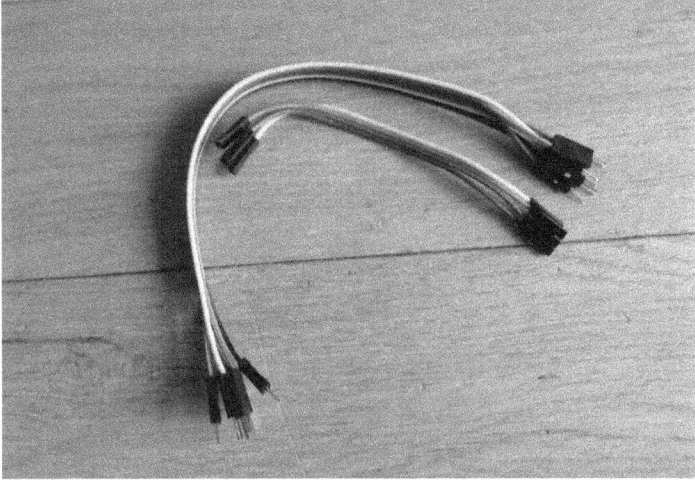

Figure 2.9: Jumper wires

Figure 2.10: Crocodile clips

Figure 2.11: Multimeter

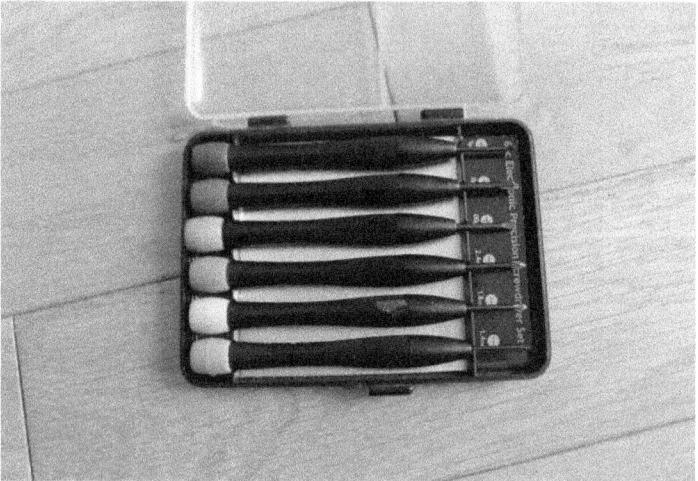

Figure 2.12: Screwdriver

Figure 2.13: Mobile Phone

Figure 2.14: Solder Iron

Figure 2.15: Solder Wire

Figure 2.16: Front Wheel

Figure 2.17: Big Back Wheel

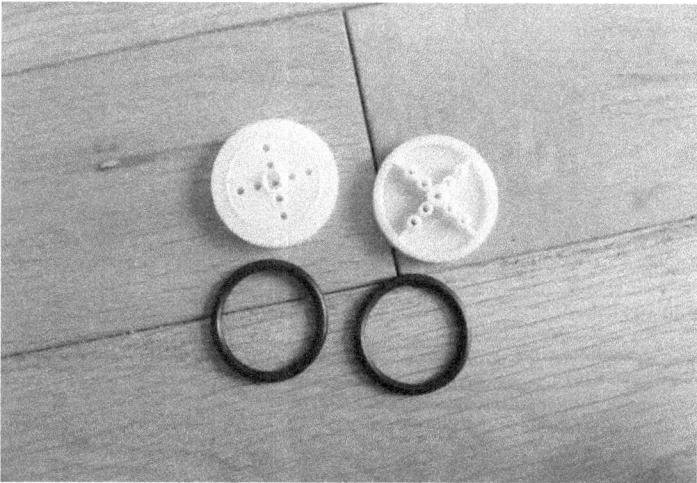

Figure 2.18: Small Wheels

Figure 2.19: Motors Fixtures

Figure 2.20: Front Wheel Fixtures

Figure 2.21: Switch

Figure 2.22: Battery Hoder

Figure 2.23: Motor Lock Wheel

Figure 2.24: Two Ways Screwdriver

Figure 2.25: Scissor

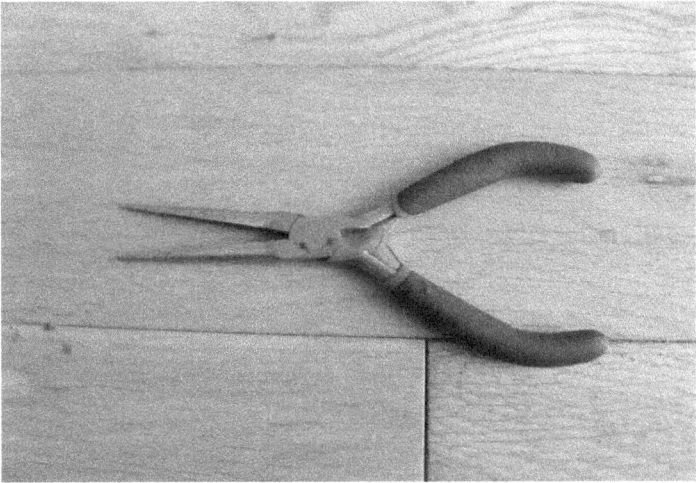

Figure 2.26: Mini Needle Nose Plier

Figure 2.27: Insulation tape

CHAPTER 3

Arduino Code

Arduino Remote Bluetooth Control

```
char t;

void setup() {

pinMode(9,OUTPUT);   //left motors forward

pinMode(10,OUTPUT);   //left motors reverse

pinMode(11,OUTPUT);   //right motors forward

pinMode(12,OUTPUT);   //right motors reverse

Serial.begin(9600);

}

void loop() {

if(Serial.available()){

  t = Serial.read();

  Serial.println(t);

}

if(t == '1'){        //move forward(all motors rotate in
forward direction)
```

```
  digitalWrite(9,HIGH);

  digitalWrite(10,LOW);

  digitalWrite(11,HIGH);

  digitalWrite(12,LOW);

}

else if(t == '2'){     //move reverse (all motors rotate in
reverse direction)

  digitalWrite(9,LOW);

  digitalWrite(10,HIGH);

  digitalWrite(11,LOW);

  digitalWrite(12,HIGH);

}

else if(t == '3'){     //turn right (left side motors rotate in
forward direction, right side motors doesn't rotate)

  digitalWrite(9,LOW);

  digitalWrite(10,LOW);

  digitalWrite(11,HIGH);

  digitalWrite(12,LOW);

}
```

```
else if(t == '4'){     //turn left (right side motors rotate in
forward direction, left side motors doesn't rotate)

  digitalWrite(9,HIGH);

  digitalWrite(10,LOW);

  digitalWrite(11,LOW);

  digitalWrite(12,LOW);

}

else if(t == '5'){     //STOP (all motors stop)

  digitalWrite(9,LOW);

  digitalWrite(10,LOW);

  digitalWrite(11,LOW);

  digitalWrite(12,LOW);

}

delay(100);

}
```

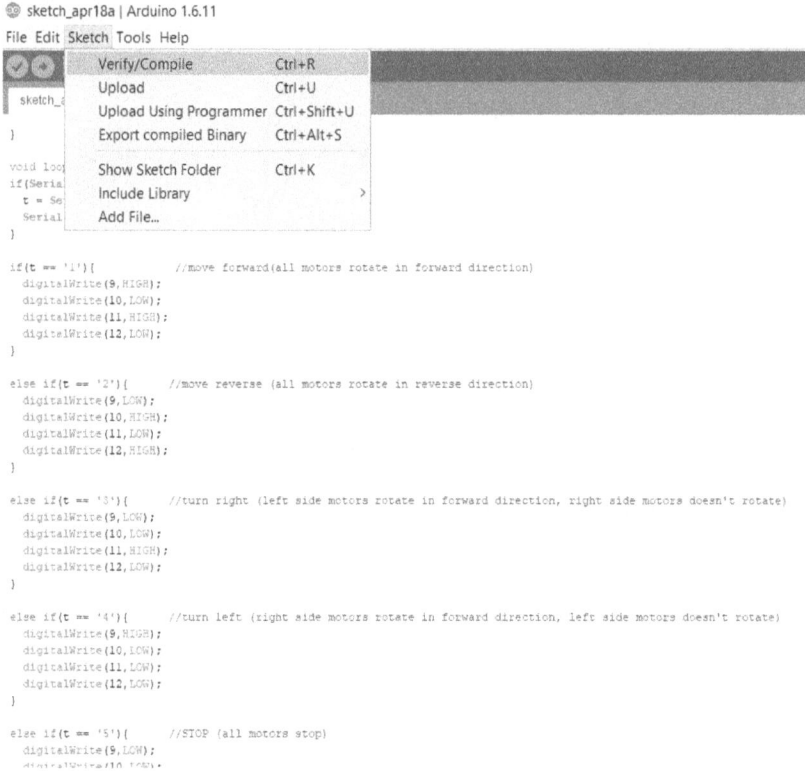

Figure 3.1:Remote Bluetooth Control

After writing the program code into the Arduino, the next step is compiled or verified.

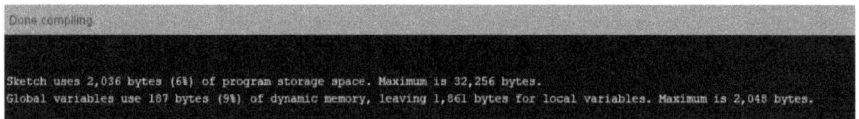

Figure 3.2: Complie

The next step is uploading to the Arduino board. At this step, we have to connect the Arduino board to the computer using the USB cable.

　　　　　　　Create Educational Robotics

Figure 3.3: Connect Arduino board to the computer

There are many kinds of Arduino boards. In this book, we would use the Arduino UNO and Arduino Mini as the main board.

At the top bar, click at the "Tools" and scroll down at the Boards Manager and choose the Arduino Board, which is connected to the computer. As the Figure 3.3, the Arduino UNO was used and connecting to the computer.

bluetooth_remote_control.ino | Arduino 1.6.11

File Edit Sketch Tools Help

bluetooth_remote

char t;

void setup() {
pinMode(9,OUTPU
pinMode(10,OUTP
pinMode(11,OUTP
pinMode(12,OUTP

Serial.begin(96(

}

void loop() {
if(Serial.avail
  t = Serial.read();
  Serial.println(t);
}

if(t == '1'){   //move forward(all motors rotate in forw
  digitalWrite(9,HIGH);
  digitalWrite(10,LOW);
  digitalWrite(11,HIGH);
  digitalWrite(12,LOW);
}

else if(t == '2'){   //move reverse (all motors rotate in reve
  digitalWrite(9,LOW);
  digitalWrite(10,HIGH);
  digitalWrite(11,LOW);
  digitalWrite(12,HIGH);
}

else if(t == '3'){   //turn right (left side motors rotate in
  digitalWrite(9,LOW);
  digitalWrite(10,LOW);
  digitalWrite(11,HIGH);
  digitalWrite(12,LOW);
}

Tools menu:
Auto Format — Ctrl+T
Archive Sketch
Fix Encoding & Reload
Serial Monitor — Ctrl+Shift+M
Serial Plotter — Ctrl+Shift+L
WiFi101 Firmware Updater
Board: "Arduino/Genuino Uno"
Port
Get Board Info
Programmer: "AVRISP mkII"
Burn Bootloader

Boards Manager...

Arduino AVR Boards
Arduino Yún
● Arduino/Genuino Uno
Arduino Duemilanove or Diecimila
Arduino Nano
Arduino/Genuino Mega or Mega 2560
Arduino Mega ADK
Arduino Leonardo
Arduino/Genuino Micro
Arduino Esplora
Arduino Mini
Arduino Ethernet
Arduino Fio
Arduino BT
LilyPad Arduino USB
LilyPad Arduino
Arduino Pro or Pro Mini
Arduino NG or older
Arduino Robot Control
Arduino Robot Motor
Arduino Gemma

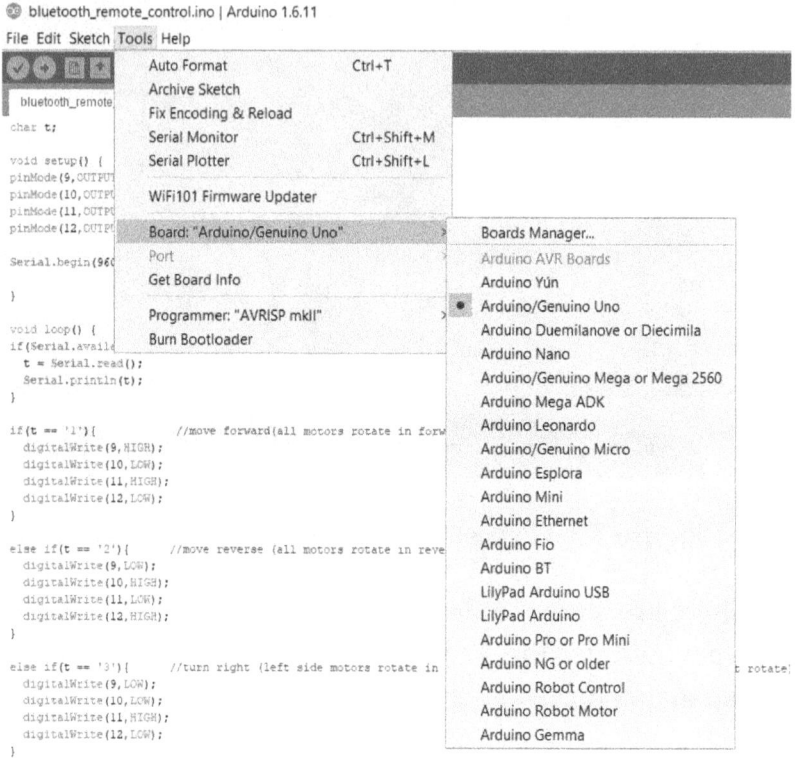

Figure 3.4: Arduino board type

Under the "Tools", scroll down to reach the Port. This is to indicate which port number in the computer that is currently connected to the Arduino board.

As shown in the Figure 3.5, the Port COM4 is connected to the Arduino UNO board.

File Edit Sketch Tools Help

Auto Format	Ctrl+T
Archive Sketch	
Fix Encoding & Reload	
Serial Monitor	Ctrl+Shift+M
Serial Plotter	Ctrl+Shift+L
WiFi101 Firmware Updater	
Board: "Arduino/Genuino Uno"	>
Port	>
Get Board Info	
Programmer: "AVRISP mkII"	>
Burn Bootloader	

Serial ports

COM4

```
char t;

void setup() {
pinMode(9,OUTPUT
pinMode(10,OUTPU
pinMode(11,OUTPU
pinMode(12,OUTPU

Serial.begin(960

}

void loop() {
if(Serial.availa
  t = Serial.read();
  Serial.println(t);
}

if(t == '1'){          //move forward(all motors rotate in forward direction)
  digitalWrite(9,HIGH);
  digitalWrite(10,LOW);
  digitalWrite(11,HIGH);
  digitalWrite(12,LOW);
}
```

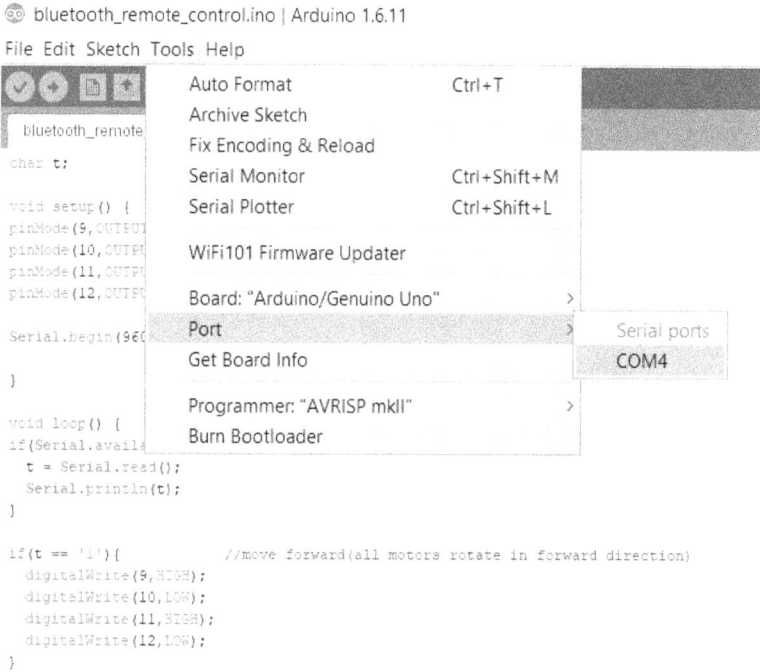

Figure 3.5: Port COM4

Next, it is ready to upload the program into the Arduino board.

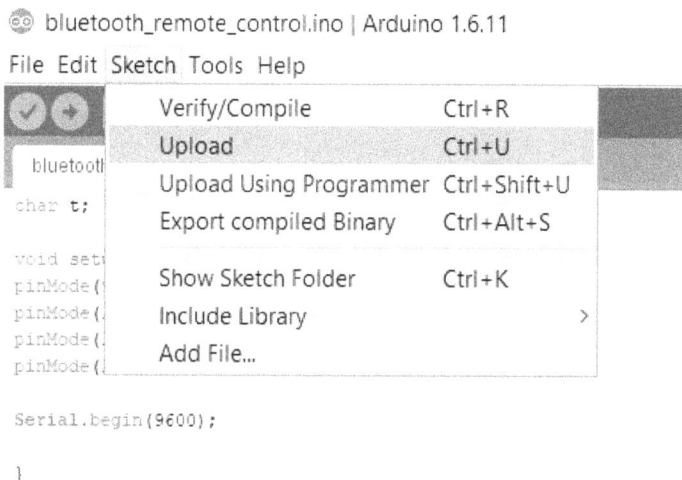

bluetooth_remote_control.ino | Arduino 1.6.11

File Edit Sketch Tools Help

Verify/Compile	Ctrl+R
Upload	Ctrl+U
Upload Using Programmer	Ctrl+Shift+U
Export compiled Binary	Ctrl+Alt+S
Show Sketch Folder	Ctrl+K
Include Library	>
Add File...	

```
char t;

void set
pinMode(
pinMode(
pinMode(
pinMode(

Serial.begin(9600);

}
```

Figure 3.6: Upload

Figure 3.7: Upload completed

Create Educational Robotics

CHAPTER 4

Motor Connection

Connection

Figure 4.1: Motors

This motor can rotate both ways, clockwise or anticlockwise. This depends upon the positive or negative voltage connect to the pole.

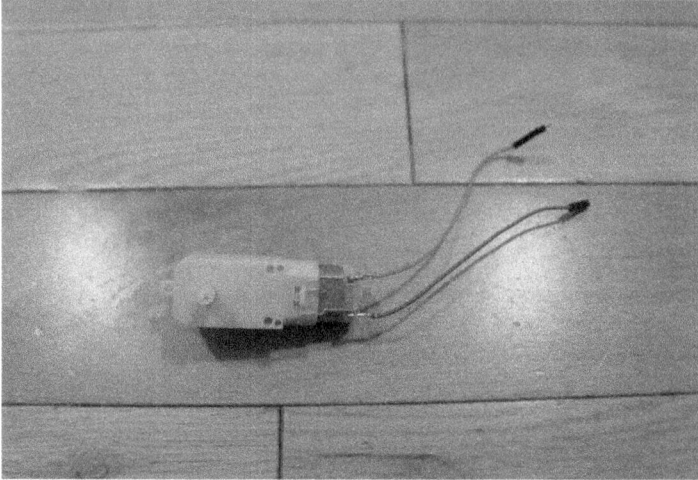

Figure 4.2: Connect cables

To make it easy to remember the direction of each motor, the different colour was used to connect to each pole in the motor.

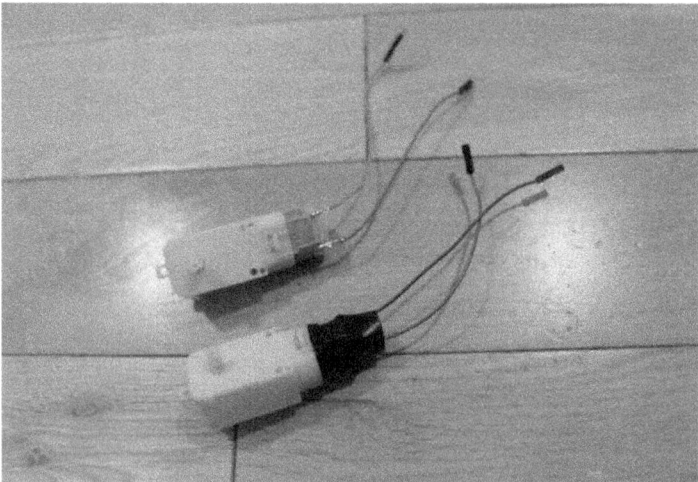

Figure 4.3: Motor cables

Create Educational Robotics

To prevent the accident of electric shock or short circuit, the insulation black tape was used to cover the connection point between the motor and the cables.

Figure 4.4: Insulation tape

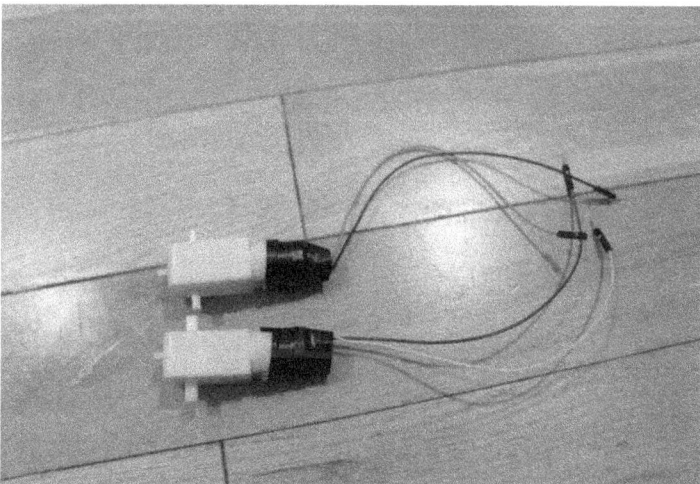

Figure 4.5: Motors with protective tape

All of the motors and motor power cables would have insulation tape cover the connection points as shown in Figure 4.5

Next, the motors need to connect to the motor controller, which this board will manage the power feed to the left motor or the right motor or both motors at the same time. Moreover, it also can change the direction of the motor by switching between positive and negative power supply.

Figure 4.6: Motor Control

The L298 Motor Drive is the motor controller. This board has a diagram as shown in Figure 4.6

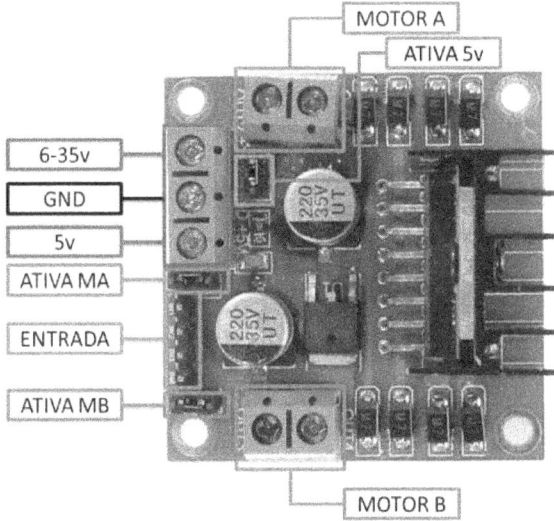

Figure 4.7: L298 Motor drive

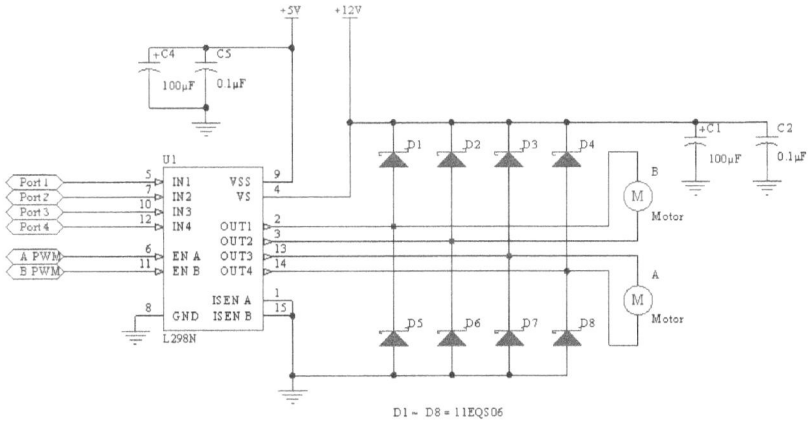

Figure 4.8: Schematic Diagram

We connect the power for motors, in this case we use more than one motor, which need more power. Thus, the connector positive would be connected into the socket 6-

35 V and the connector negative would be connected into the ground, GND.

Figure 4.9: Power supply 6-35 V

Figure 4.10: Power supply 5V

In addition, the power supply could be used the 5V and GND as shown in the Figure 4.9

Figure 4.11: Power supply 5V and 6-35 V

Figure 4.12: Power supply to Arduino

However, there are many options that we can use the 5V as the power supply for others, such as the 5V could be the power supply for the Arduino board.

Figure 4.13: Connect to the right motor

Figure 4.14: Connect to the left motor

This motor control board could be used to control individual motor and more than two motors at the same time.

Figure 4.15: two motors

Figure 4.16: Four motors

This motor control can also control the direction of each motor on the left and on the right. It could manage the direction:

- Move forward,

- Move backward,

- Both move in the same forward direction

- Both moves backward at the same direction

- The left motor working, while the right motor stop

- The right motor working, while the left motor stop

Figure 4.17: Control motor direction

The N1, N2, N3 and N4 is the port 1 to port 4 to control each side of the motors to act as command as the program code indicated.

Figure 4.18: Power supply and motor control

Figure 4.16 shows the power supply to the motor via the socket 6-35 V and 5V,the control motor port N1, N2, N3 and N4.

Figure 4.19: One motor connects to motor control

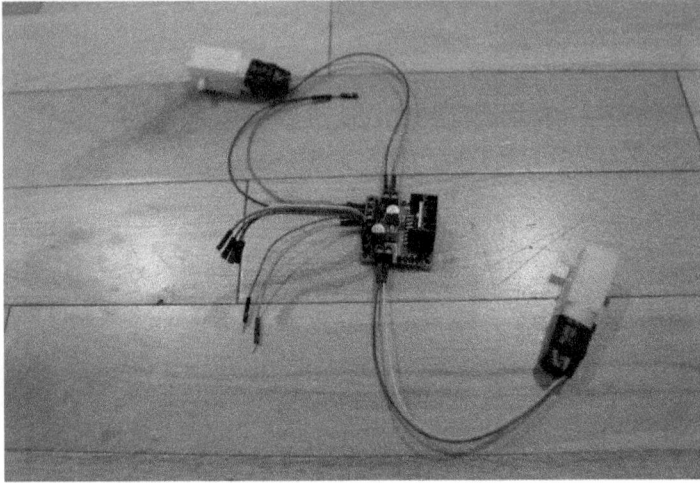

Figure 4.20: Two motors connect to motor control

CHAPTER 5

Arduino Connection

Arduino Connection

Figure 5.1: Arduino board

The next step is to connect the L298 motor control, which is shown in the figure 5.2 into the Arduino UNO board in the Figure 5.1

There are options to feed 5V power supply from the L298 motor control board directly into the Vin in the ArduinoUNO board.

In addition, the main functions to control the motors is the ENTRADA as shown in the Figure 5.3, the L298 Motor Drive is the motor controller.diagram as shown in Figure 5.3

Figure 5.2: Motor Control

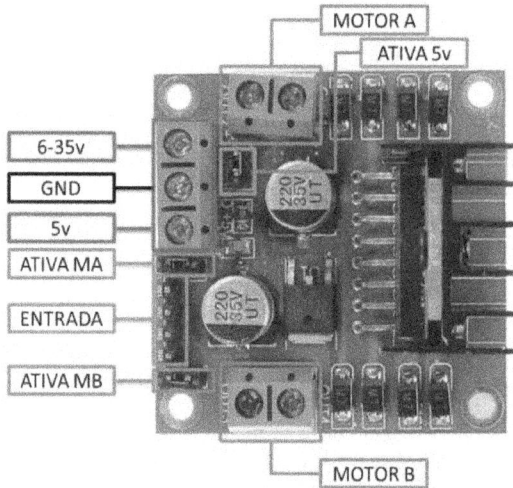

Figure 5.3: L298 Motor drive

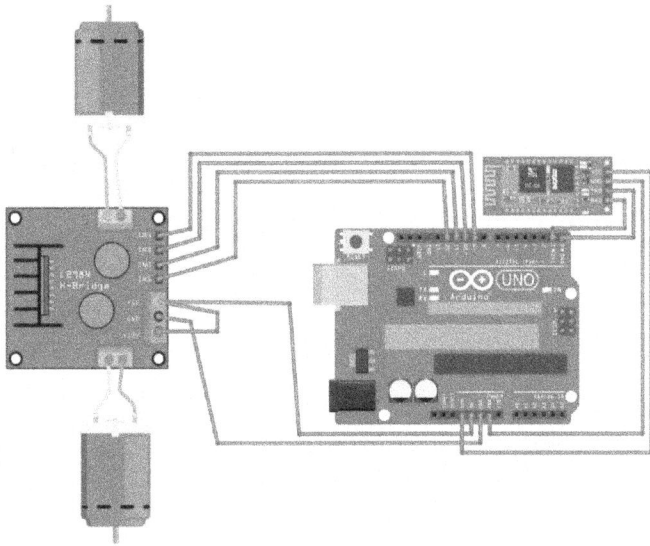

Figure 5.4: Circuit diagram

Figure 5.5: N1-N4 Connector

The connector port IN1 to IN4 from the motor control connect to the Arduino UNO board as follows:

- IN 1 connect to pin 9

- IN2 connect to pin 10

- IN3 connect to pin 11

- IN4 connect to pin 12

Figure 5.6: Control motor from Arduino

This is the control motors from the program in the Arduino board. The connectors from the IN1, IN2, IN3 and IN4 to the Arduino UNO board at the pin number 9, 10, 11 and 12 as shown in the Figure 5.6

Figure 5.7: Power from control to Arduino board

There is an option to get power supply 5V from the motor control board feed to the Vin and GND as shown in Figure 5.7

Figure 5.8: HC-05 Bluetooth Module

One more equipment that we need to perform actions via the remote control using the HC-05 Bluetooth Module.

This would allow us to use the mobile phone to control the direction of the motors.

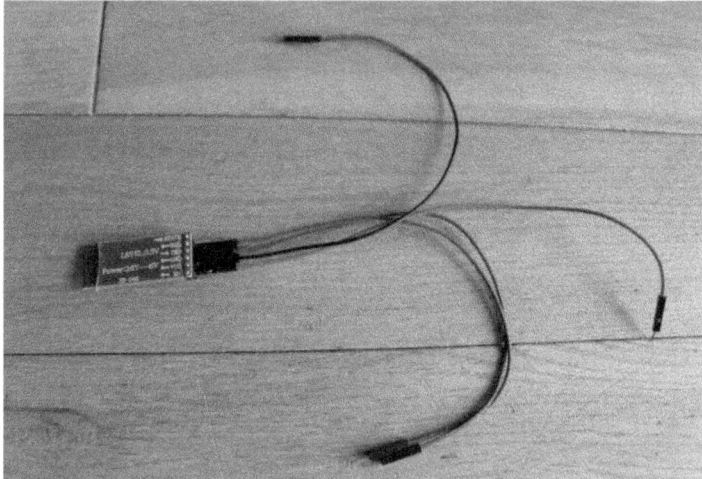

Figure 5.9: Bluetooth Connection

There are four connectors between the Bluetooth and Arduino UNO, they are as follows:

Vcc Bluetooth connects to Arduino at pin 5V

GND Bluetooth connects to Arduino at pin GND

TXD Bluetooth connects to Arduino at pin RX

RX Bluetooth connects to Arduino at pin TX

Figure 5.10 shows the cable connection between the HC-05 Bluetooth and the Arduino Bluetooth.

Figure 5.10: Bluetooth connects to Arduino

Figure 5.11: Power supply

The power supply for the Bluetooth is 5V. Therefore, we can use the power feeder 5V from the motor control board as shown in the figure 5.11

Figure 5.12: Bluetooth, motor control and Arduino

Figure 5.12 shows the whole connection with the HC5 Bluetooth, L298 Motor control and Arduino UNO. The program has to upload into the Arduino UNO before connecting with any board.

Next, we need to connect two motors into the motor control.

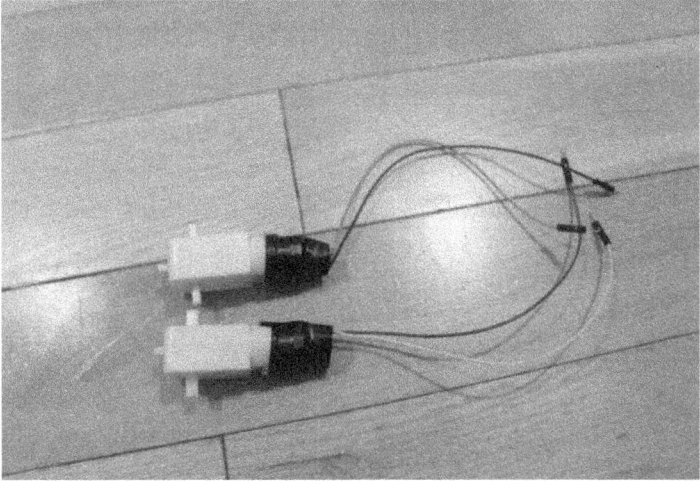

Figure 5.13: Two motors

Figure 5.14: Control motor board and motors

Figure 5.14 shows two motors connect to the L298 Motor control, one motor on the left and one motor on the right. There are two types of wheels, the big wheel and small wheel, as shown in Figure 5.15 and Figure 5.16

Figure 5.15: Big wheels

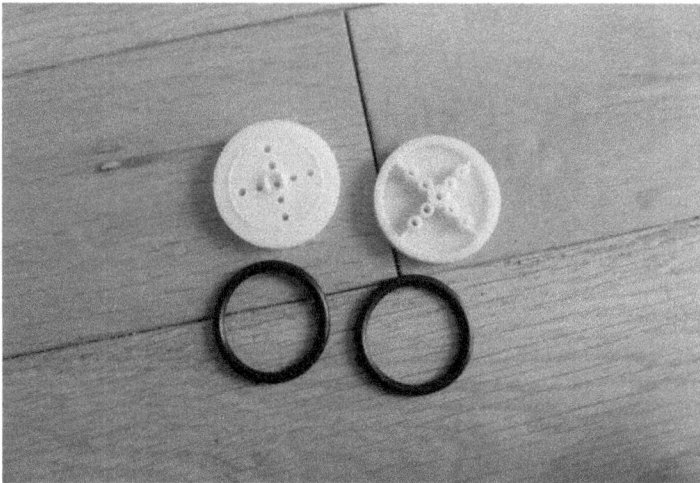

Figure 5.16: Small wheels

The small wheel comes as two parts, the rubber circle and the centre part. The rubber needs to strength up and get into the slot of the centre part.

Figure 5.17: Left motor with the small wheel

Figure 5.18: Left and Right motors with small wheels

Figure 5.19: Right motor with big wheel

Figure 5.20: Left and Right motors with big wheels

All the connectors on the motor control board are shown in Figure 5.21

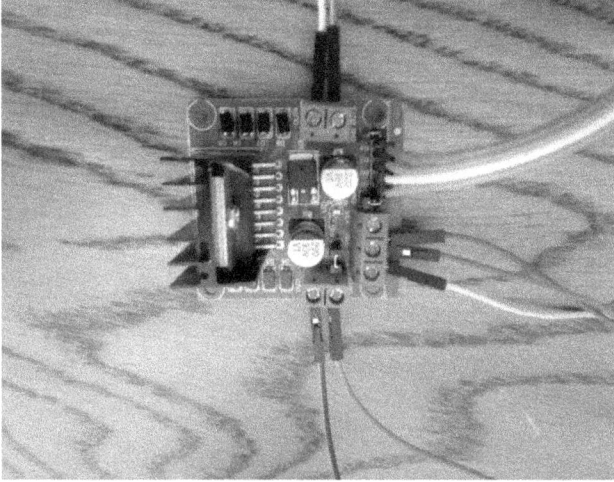

Figure 5.21: Connectors on the motor control

There seven cables connect to the control motor board. There four cables for two motors, two cables for positive and two for negative. Four cables for IN1 to IN4 for control the motor directions and running.

In addition, there are two cables as a power supply for motors, which connect to the socket 6-35 V and GND. Also, the 5V to feed the power to the Arduino UNO board at the pin Vcc.

Figure 5.22 shows the connectors on the Arduino UNO board. There are four cables from the Bluetooth, which connect to 5V, GND, TX and RX as shown in Figure 5.23 Furthermore, there are two cables connect at pin Vcc and GND from the control motor board.

Figure 5.22: Connectors on the Arduino UNO

Figure 5.23: Connectors on the Bluetooth

CHAPTER 6

Testing

Test Connection

Figure 6.1: Power adapter

It is a useful method to use the power adapter as power supply to the motors. This is because as the amount of motors increase, they would need more energy.

This power adapter can be adjusted from 3 V, 4.5 V, 6 V, 7.5 V, 9 V and 12 V. The positive connects to the socket 6-35V and the negative connect to socket GND on the motor control board.

Figure 6.2: Power supply to 6-35V

Another power supply is the battery set is using the AA, 1.5 V parallel four sets, it would be 6 V, as shown in Figure 6.3

Figure 6.3: AA Battery set

Figure 6.4: Vcc Battery set supply to Arduino

The four battery set is suitable for a small load and need power no more than 6 V. For example, the Bluetooth board needs 5V power supply.

One more power supply for the Arduino board is via the USB cable, as shown in Figure 6.5, where the USB would connect directly into the computer. This also is the power supply for the Bluetooth board.

Therefore, we use the power adapter as the power supply for the motor and the four AA battery set or USB cable as the power supply for the Arduino board and the Bluetooth board.

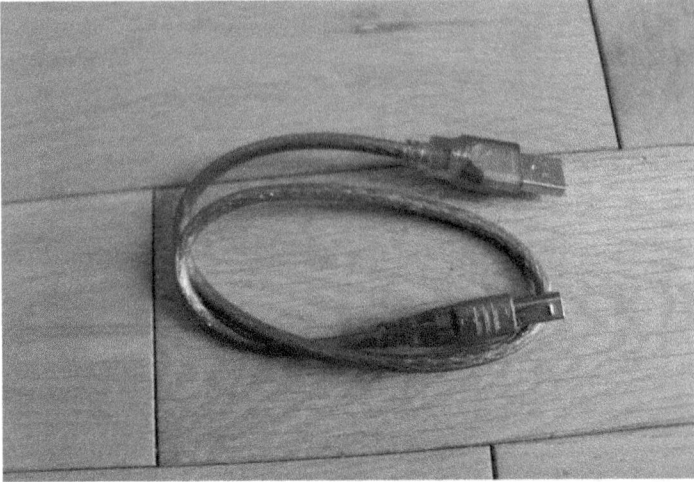

Figure 6.5: USB cable

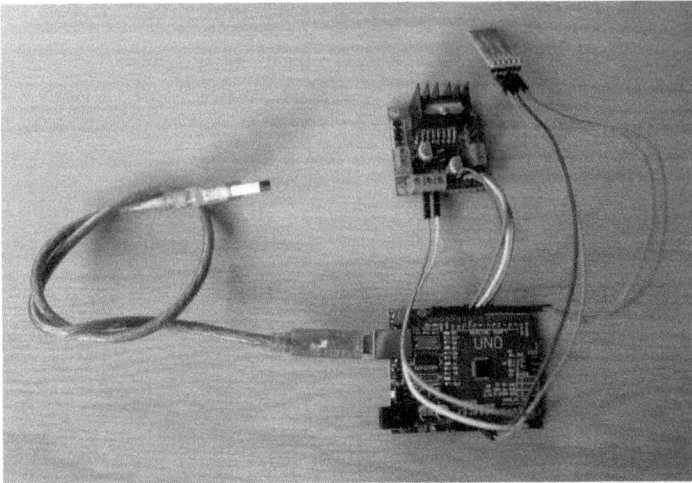

Figure 6.6: USB power supply to Arduino board

To make the circuit check easily, the power supply for the motors is separated from the power supply to the Arduino UNO board.

The multimeter would use to check the circuit such as the voltage. Also, it needs to check the current and resistance at each point in the circuit.

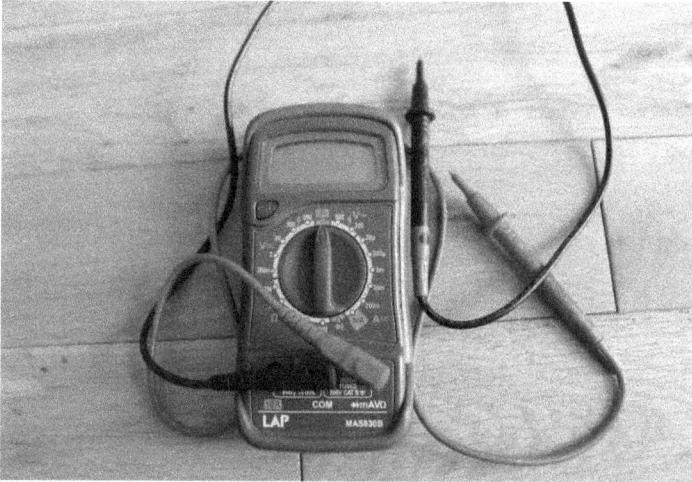

Figure 6.7: Multimeter

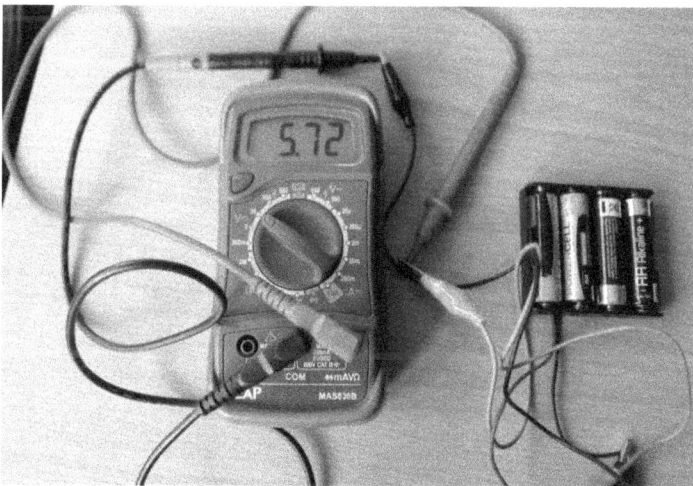

Figure 6.8: Using the Multimeter

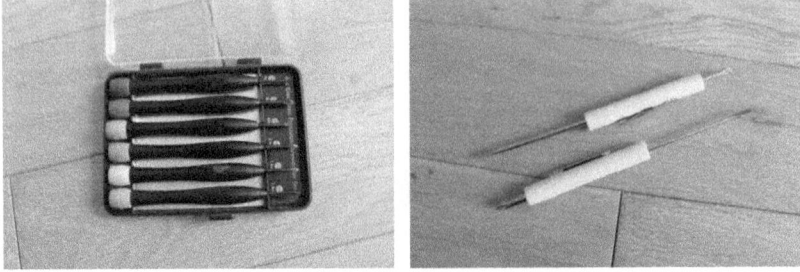

Figure 6.9: Screwdriver

To build the robotics, we need the screwdriver to build the body and connect the cables and connectors between the motors and the motor control board, the motor control board and the Arduino board and power supply.

Figure 6.10: Using the screwdriver

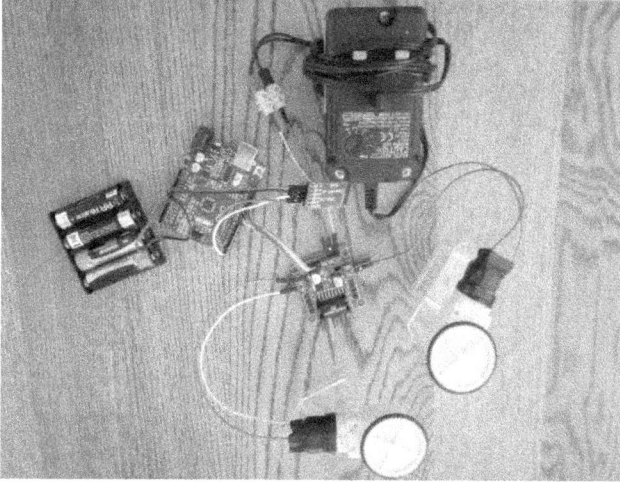

Figure 6.11: All components

Figure 6.11 shows the all components to do the test drive the car robot. The program was uploaded into the Arduino was shown in Chapter 2.

The Bluetooth would run as the motor control to drive the motors. While the signals would receive from the Bluetooth from the mobile phone.

Figure 6.12: Mobile phone

CHAPTER 7

Mobile App

Install Mobile App

Figure 7.1: Install mobile app

At the play Store on your mobile phone, search for the Arduino Bluetooth control.

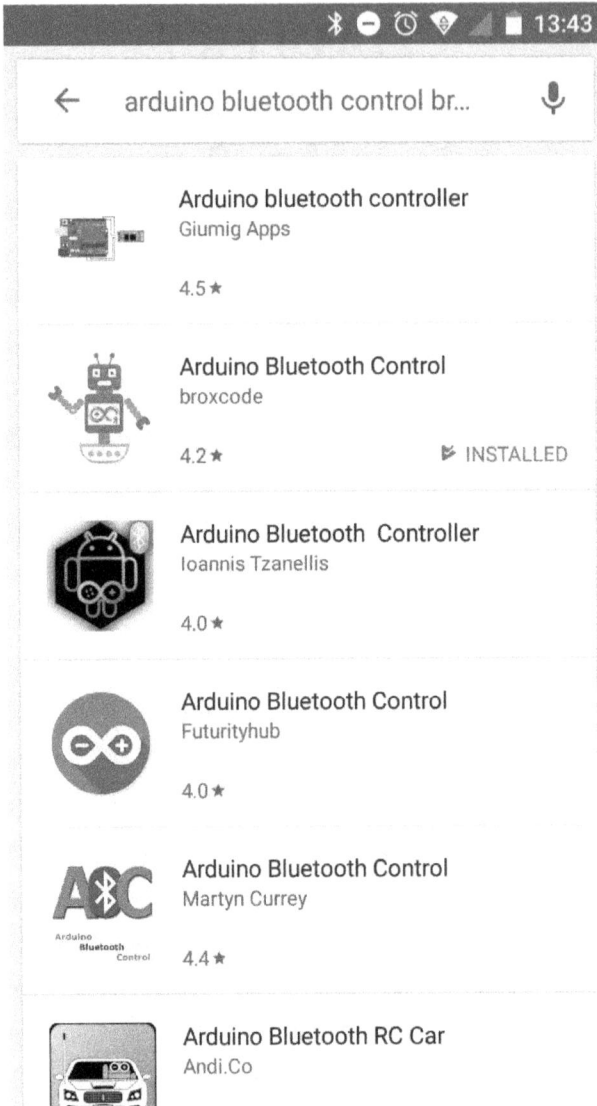

Figure 7.2: Arduino Bluetooth control

We choose to install the Arduino Bluetooth Control by Broxcode.

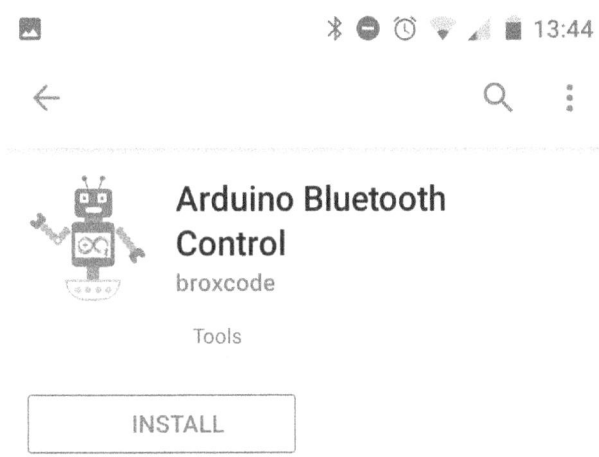

← 　　　　　　 🔍 　 ⋮

Arduino Bluetooth Control

broxcode

Tools

INSTALL

What's new ●

Last updated 23 Mar 2019

SMS feature no longer supported due to Google policy update.

Exciting new features to be coming very soon ! St

READ MORE

Rate this app

Tell others what you think

☆ 　 ☆ 　 ☆ 　 ☆ 　 ☆

Write a review

Figure 7.3: Install Arduino control App

Click Install and we would have the icon on the mobile phone as shown in Figure 7.4

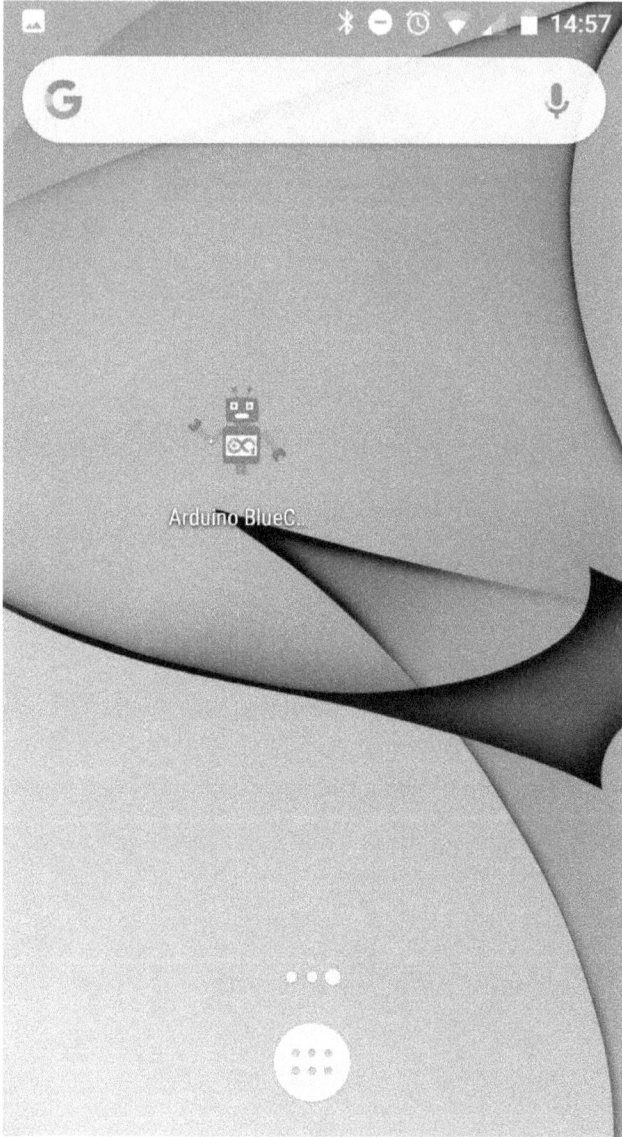

Figure 7.4: App Arduino control icon

When the Arduino control App starts, there are many options as a control system as shown in Figure 7.5

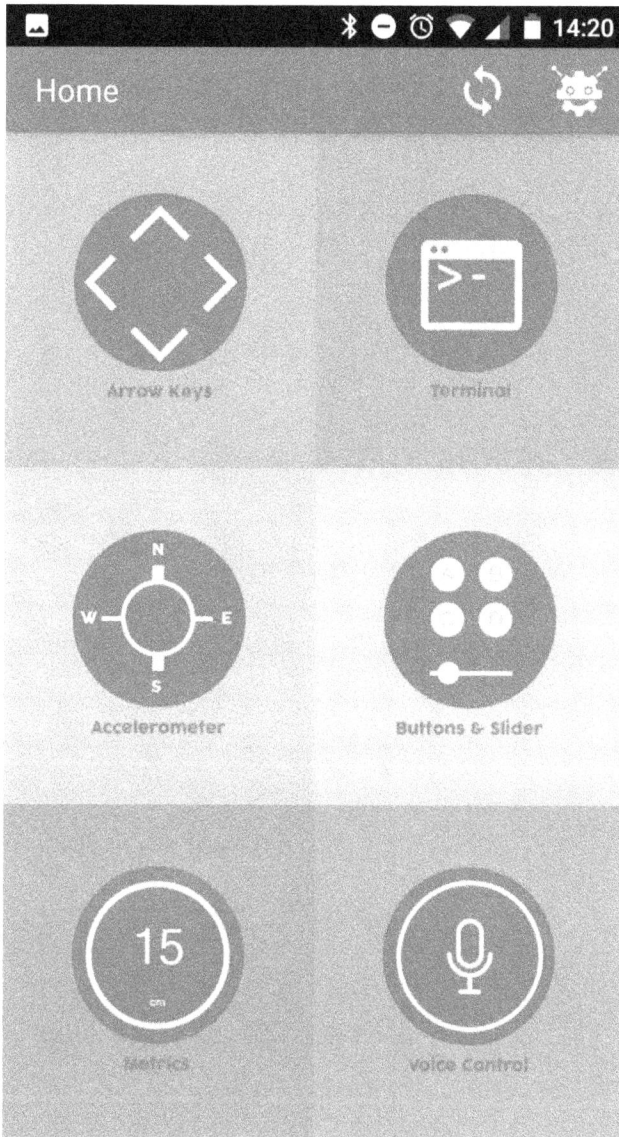

Figure 7.5: Arduino Control Screen

Next we need to connect the Bluetooth from a mobile phone to the Arduino board and HC-05 Bluetooth. Click on the arrow circle button as the red indication is shown in the Figure 7.6

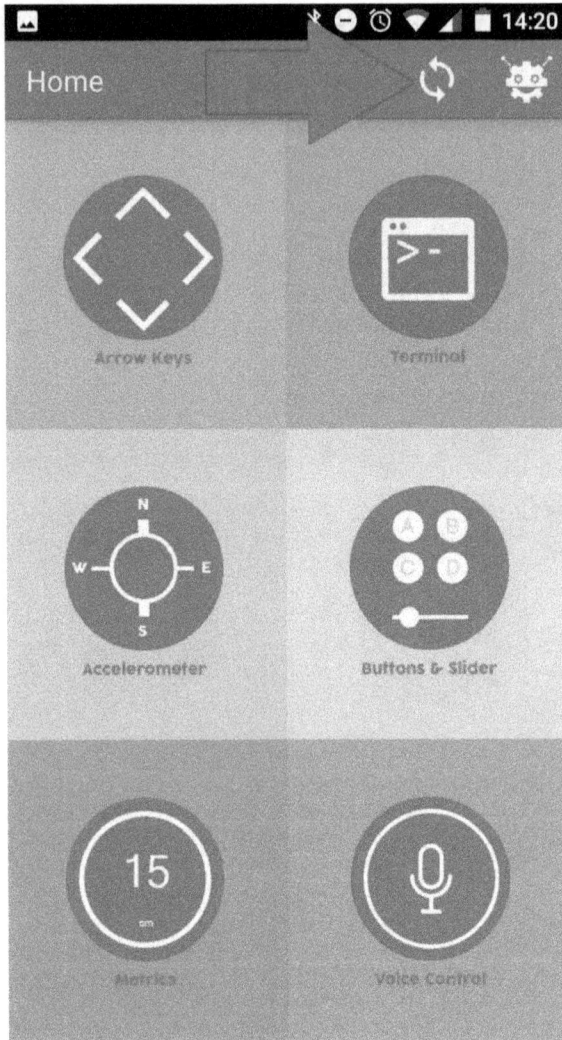

Figure 7.6: Connect Bluetooth

The HC-05 is our Arduino Bluetooth device to connect
Bluetooth to the mobile phone.

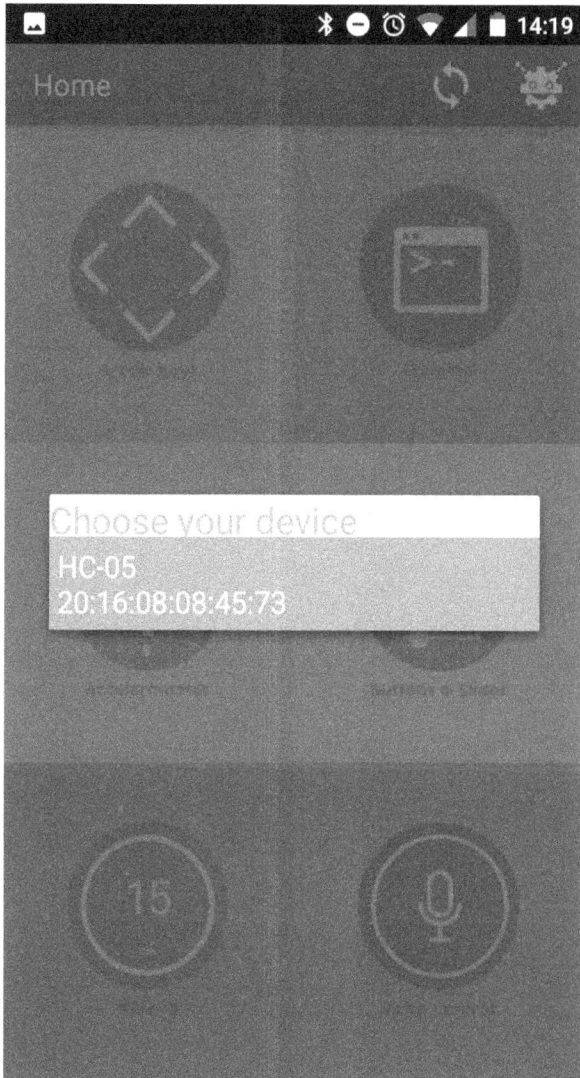

Figure 7.7: Choose HC-05 Bluetooth

The Bluetooth mobile will start to connect to the Bluetooth Arduino.

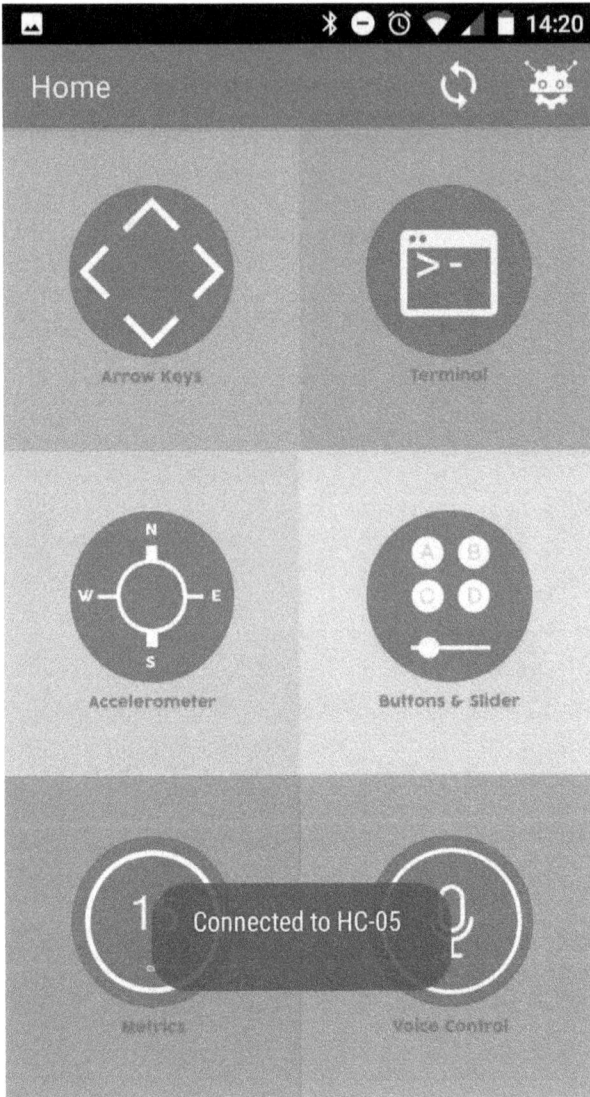

Figure 7.8: Connect to HC-05 Bluetooth

If the Bluetooth from mobile phone cannot connect to Bluetooth on the Arduino and HC-5 Bluetooth. The screen shows as in Figure 7.9

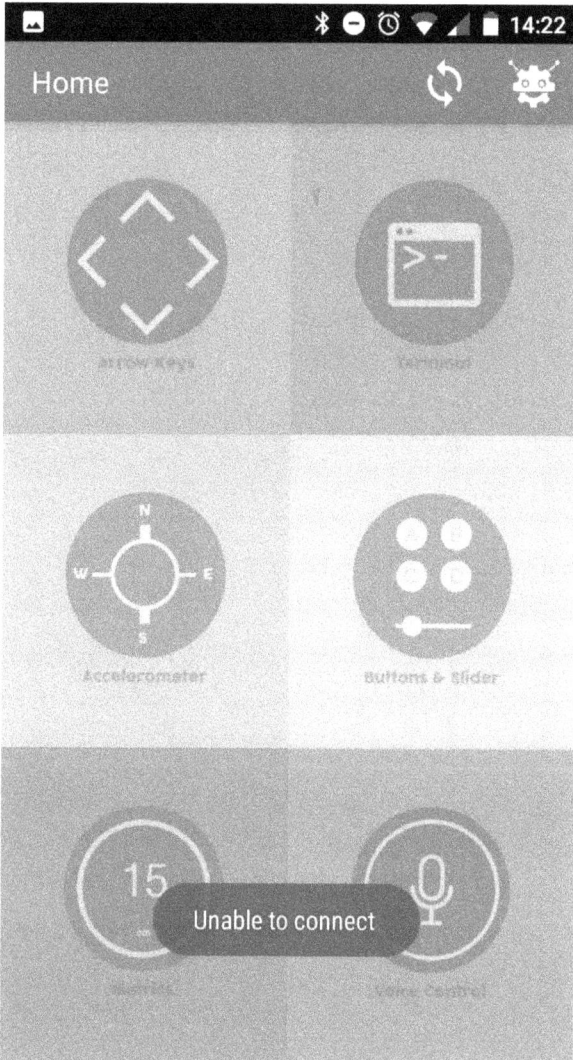

Figure 7.9: Unable to connect

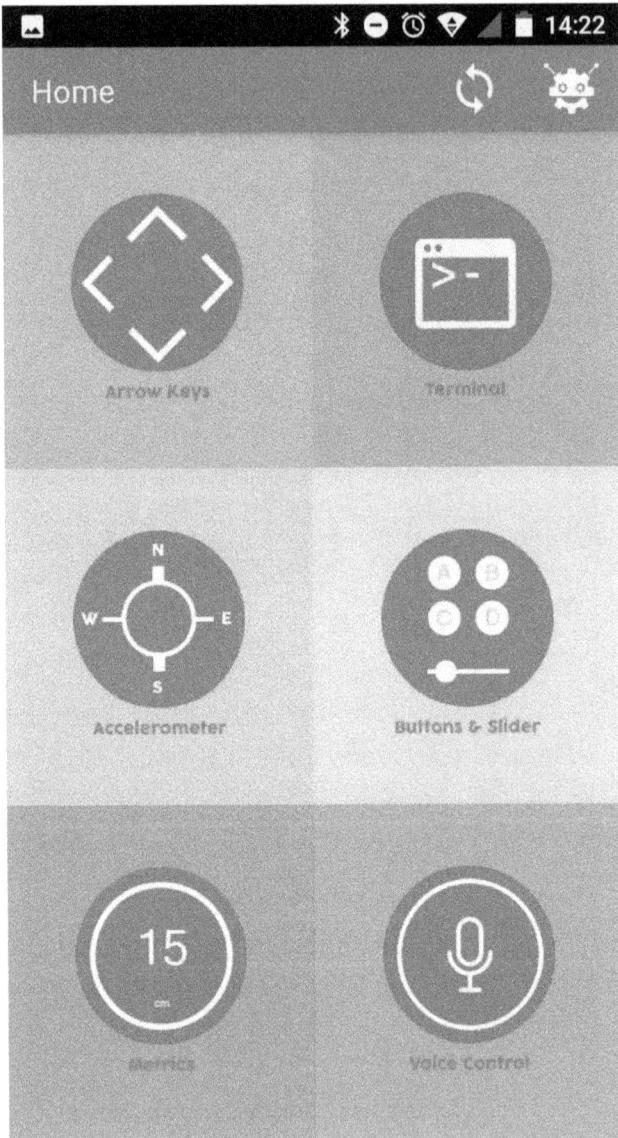

Figure 7.10: Arduino control options

There are six functions for this Arduino control App. There are Arrow Keys, Terminal, Accelerometer, Button & Slide, Metrics and Voice Control, as shown in Figure 7.11-16

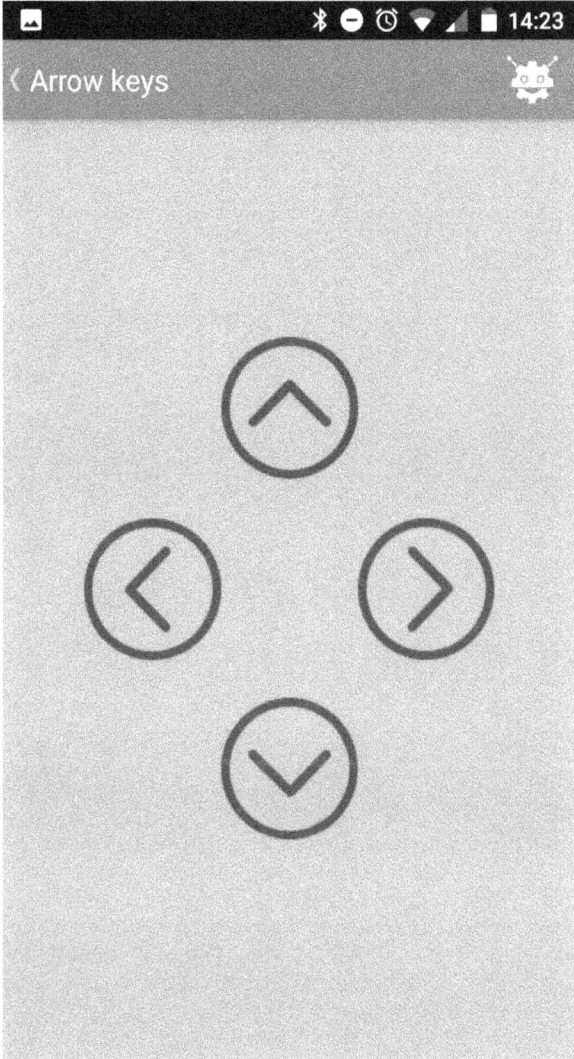

Figure 7.11: Arrow keys

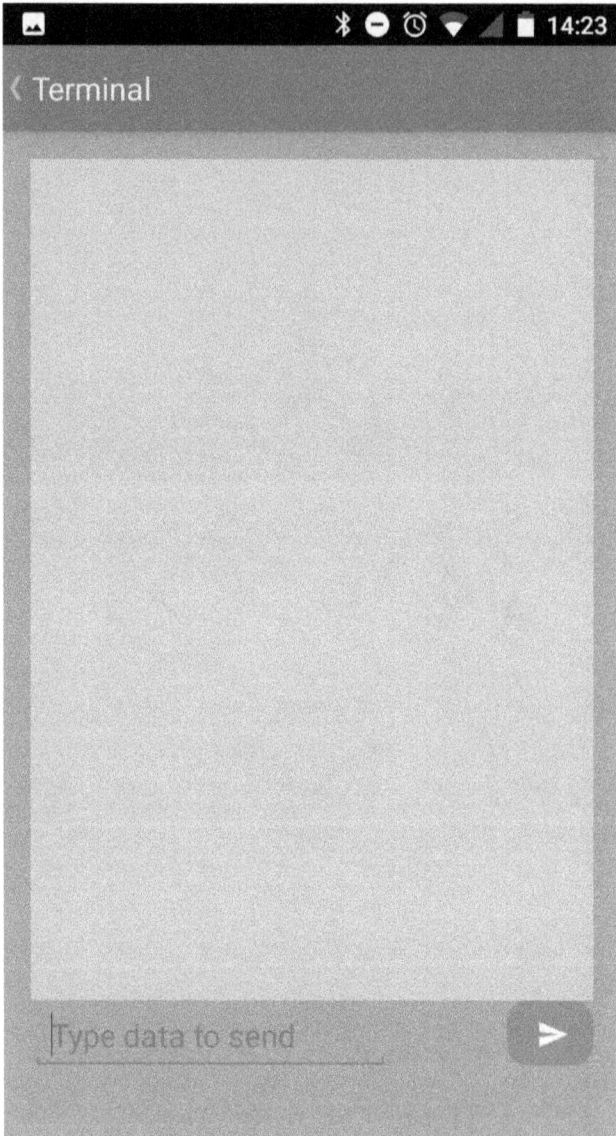

Figure 7.12: Terminal

Create Educational Robotics

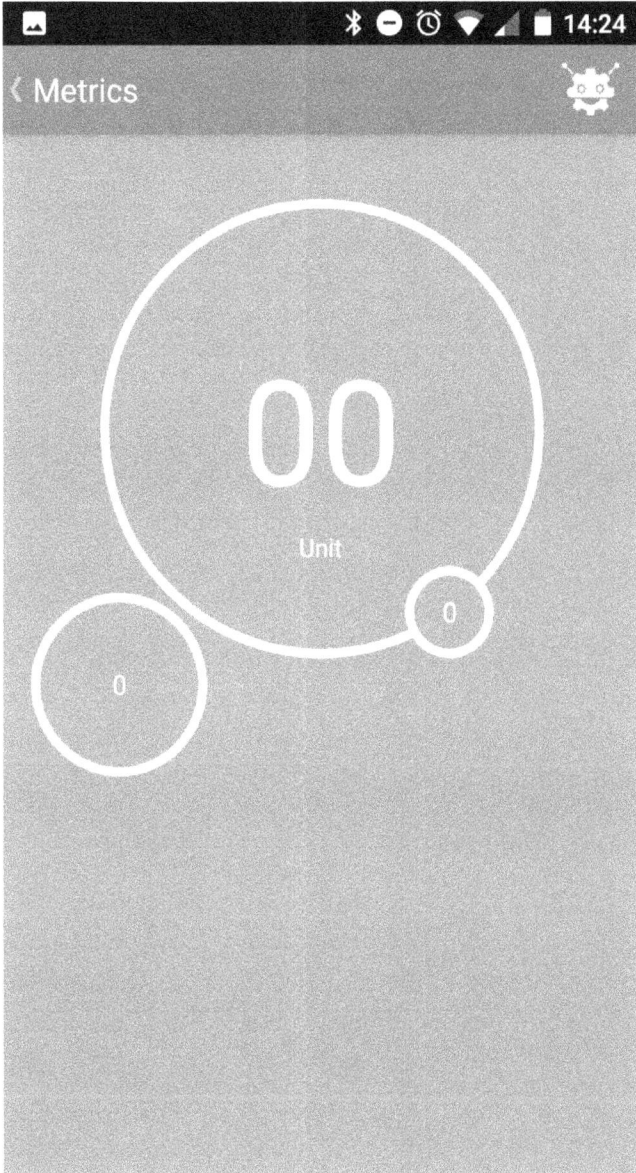

Figure 7.13: Metrics

Figure 7.14: Voice Control

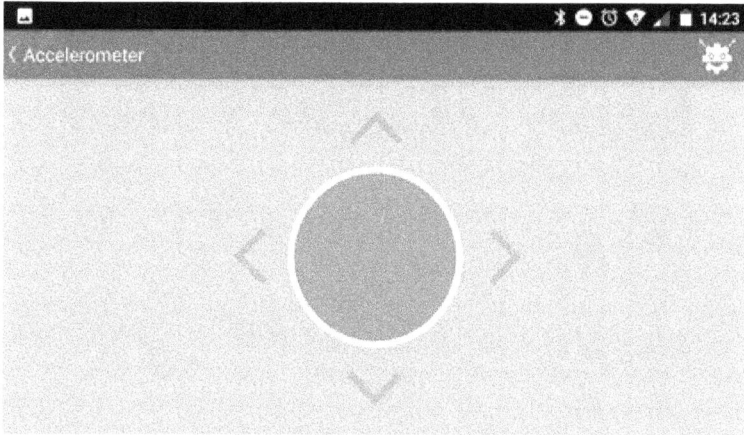

Figure 7.15: Accelerometer

In this book, we will use the Buttons function to control the Robotic Car, as shown in Figure 7.16

Button 1 (A): Left motors and Right motors move forward

Button 2 (B): Left motors and Right motors move backward

Button 3 (C): Left motors work

Button 4 (D): Right motors work

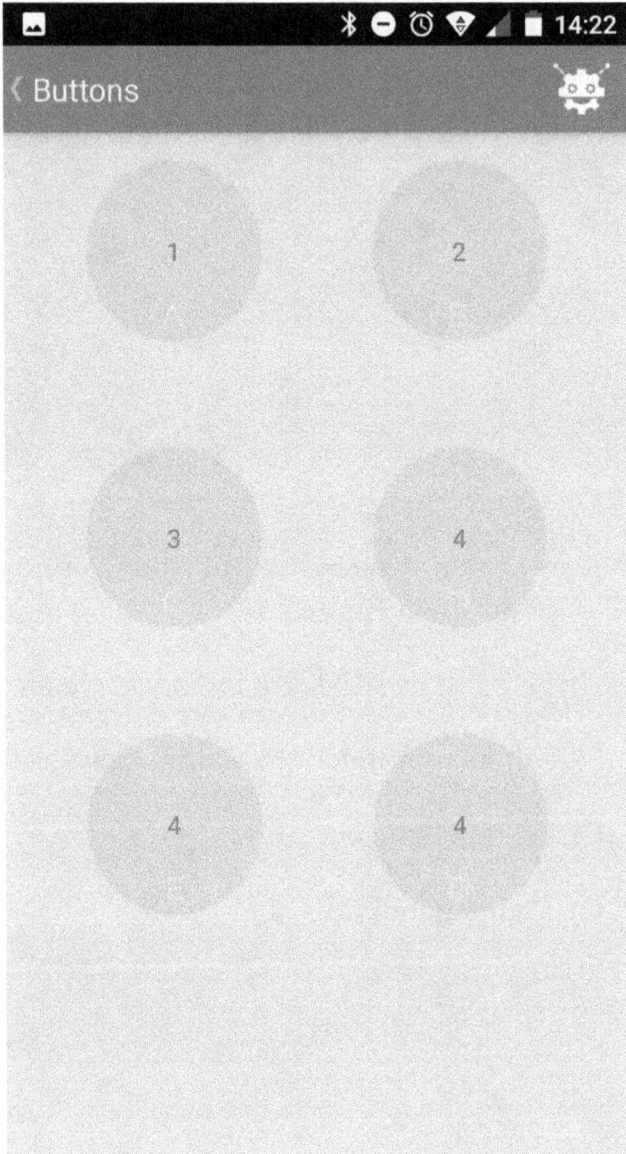

Figure 7.16: Buttons

CHAPTER 8

Build the Body

Build the Body

There are many shapes and sizes of the robotics car bodies.

A. Robotic Four Wheels

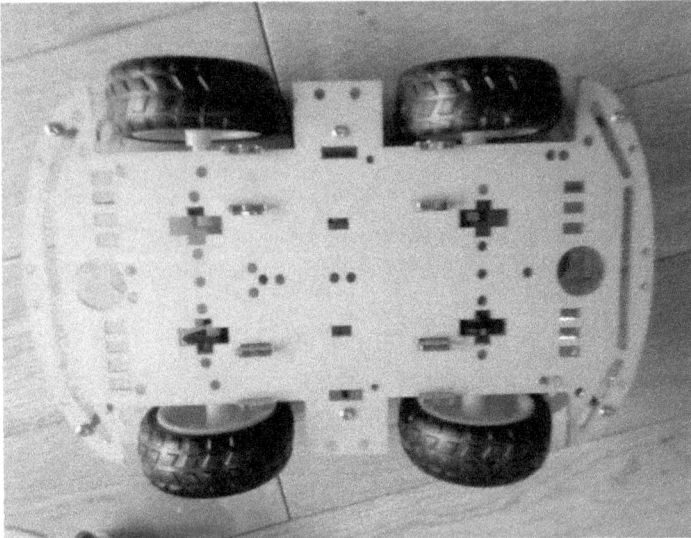

Figure 8.1: Robotic Four Wheels

B. Robotic Three Wheels

Figure 8.2: Robotic Three Wheels

C. Robotics Smile Eyes

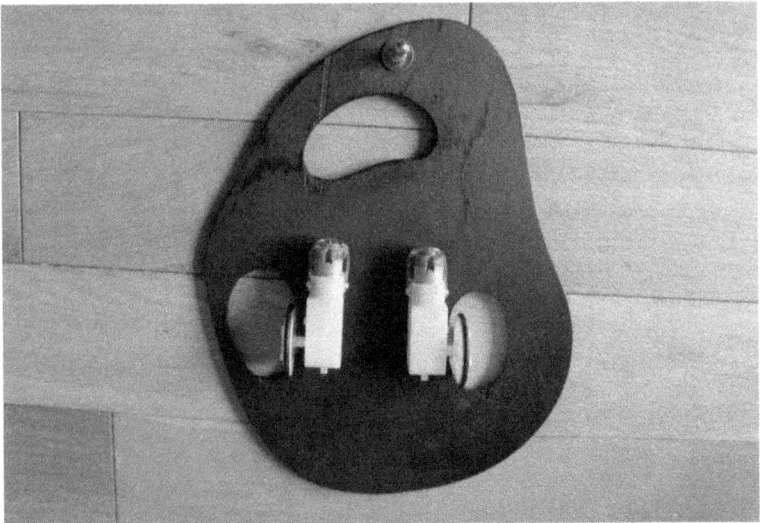

Figure 8.3: Robotics Smile Eyes

D. Robotics Five Toes Foot Pain wood

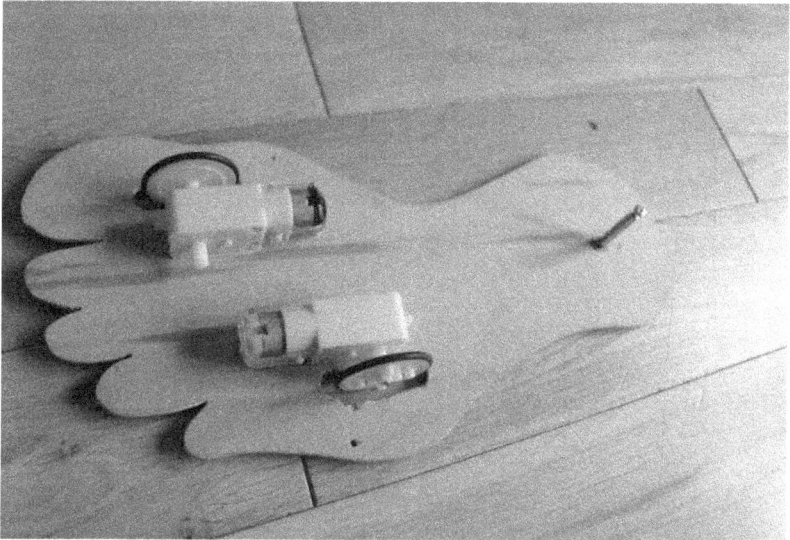

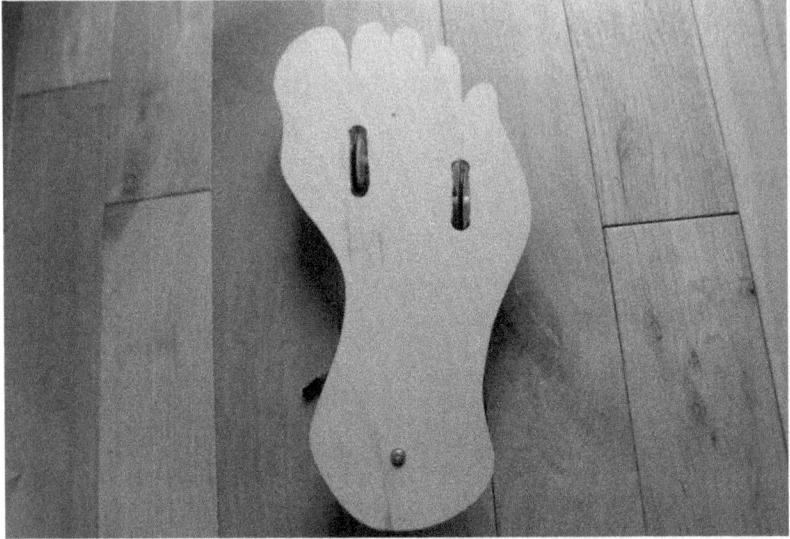

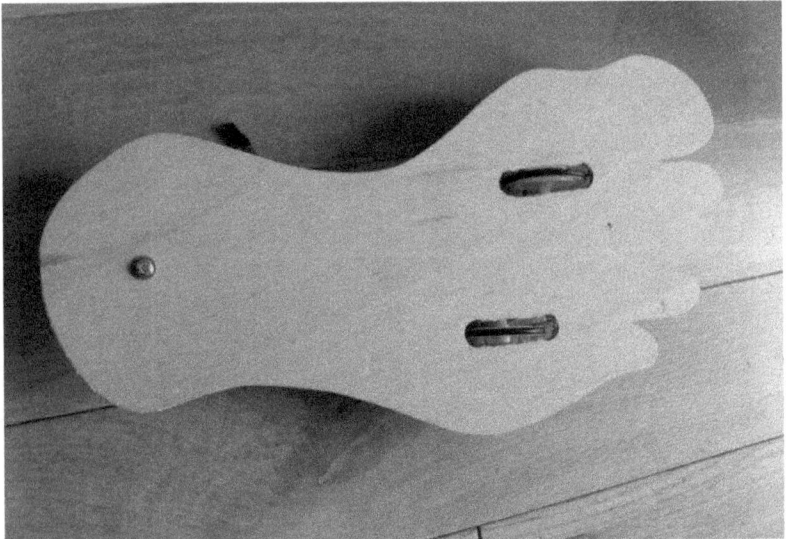

Figure 8.4: Robotics Five Toes Foot Pain wood

E. Robotics Five Toes Foot colour lines

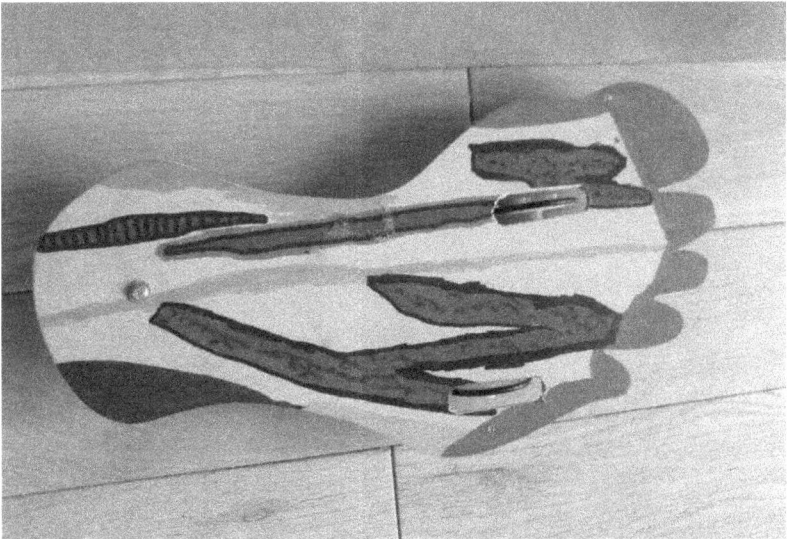

Figure 8.5: Robotics Five Toes Foot colour lines

F. Robotics Five Toes Foot colour dots

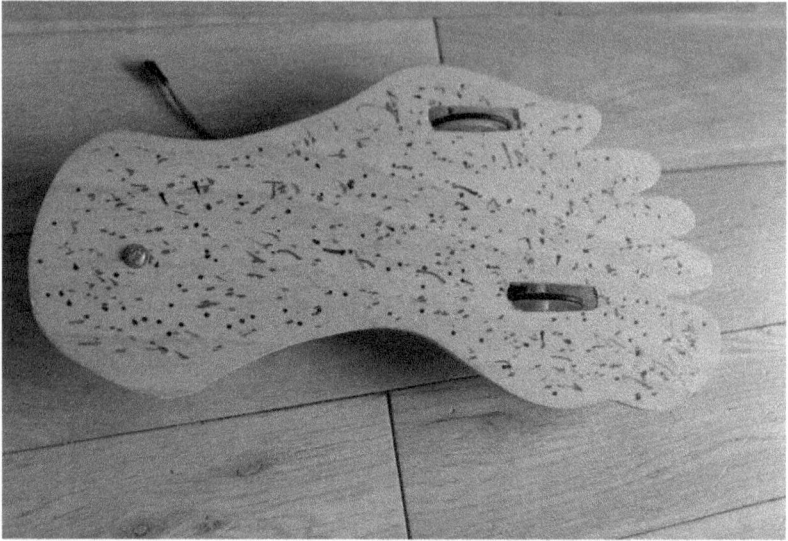

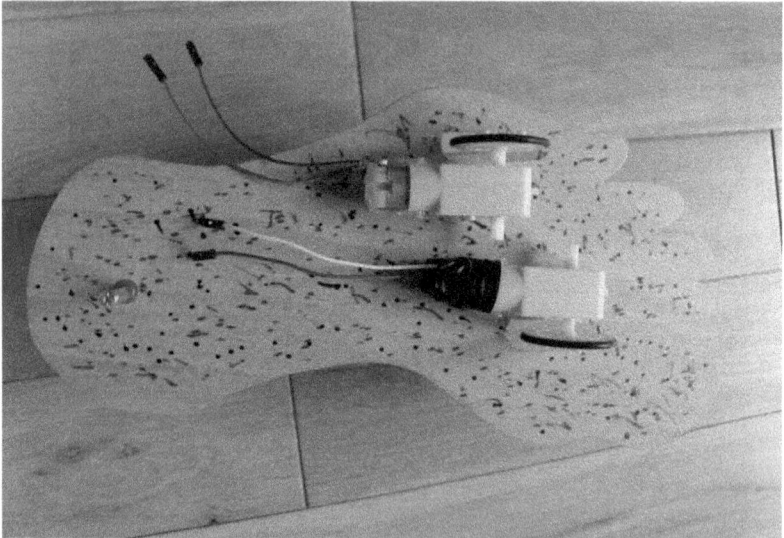

Figure 8.6: Robotics Five Toes Foot colour dots

G. Robotics Five Toes Foot colour groups

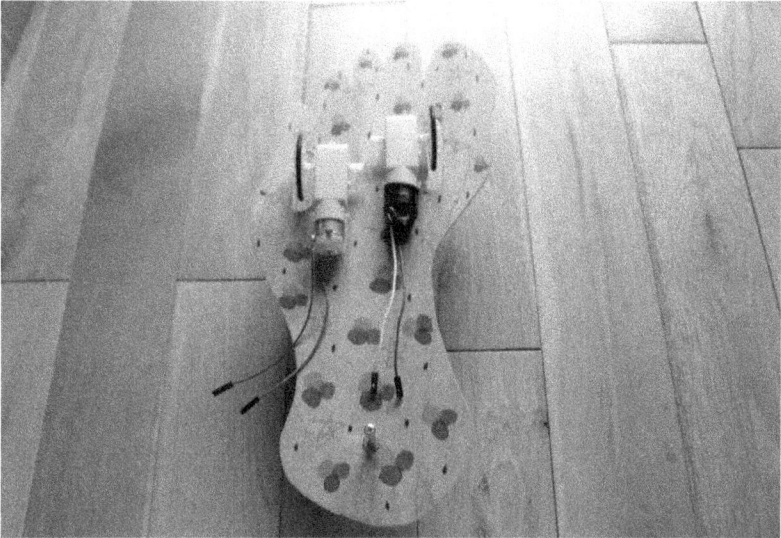

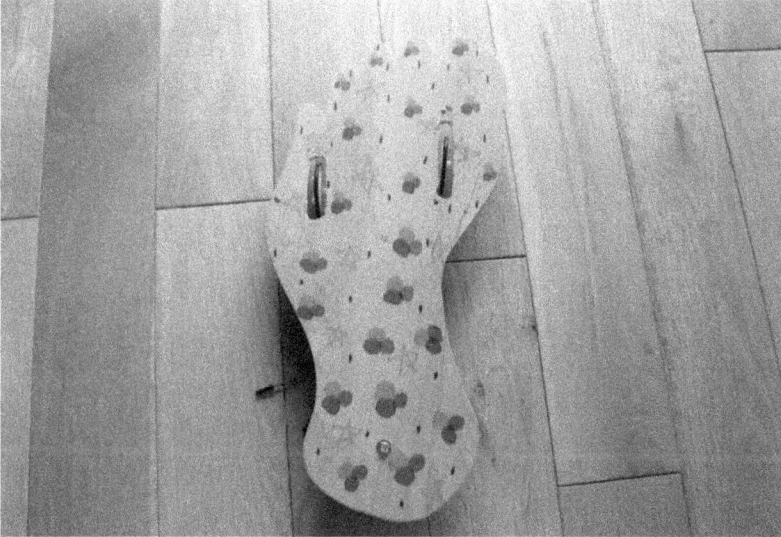

Figure 8.7: Robotics Five Toes Foot colour groups

H. Robotics Five Toes Foot colour lines and dots

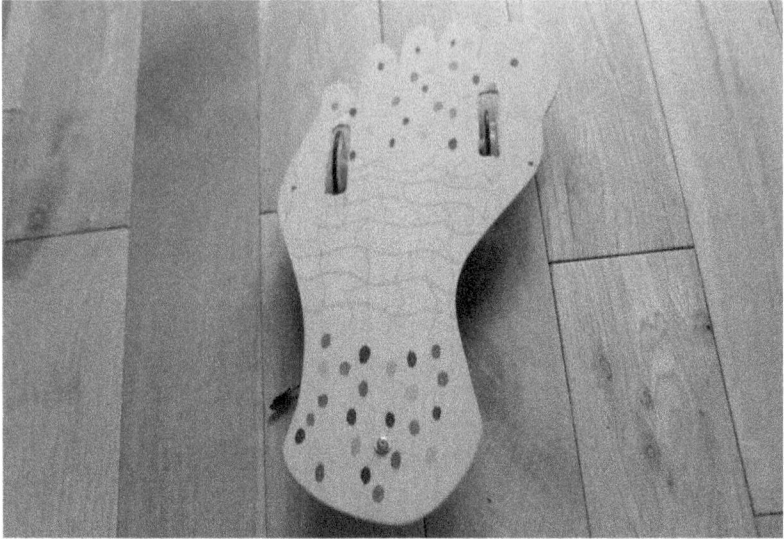

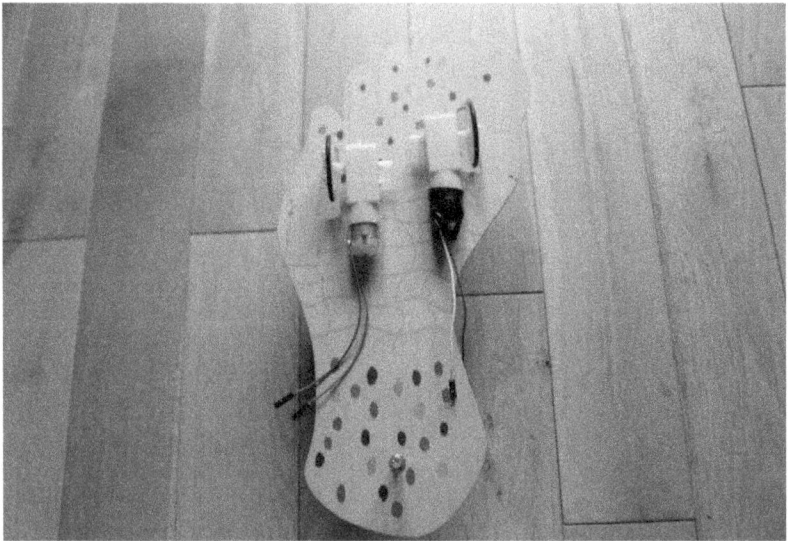

Figure 8.8: Robotics Five Toes Foot colour lines and dots

I. Robotics Five Toes Foot Baby Jesus

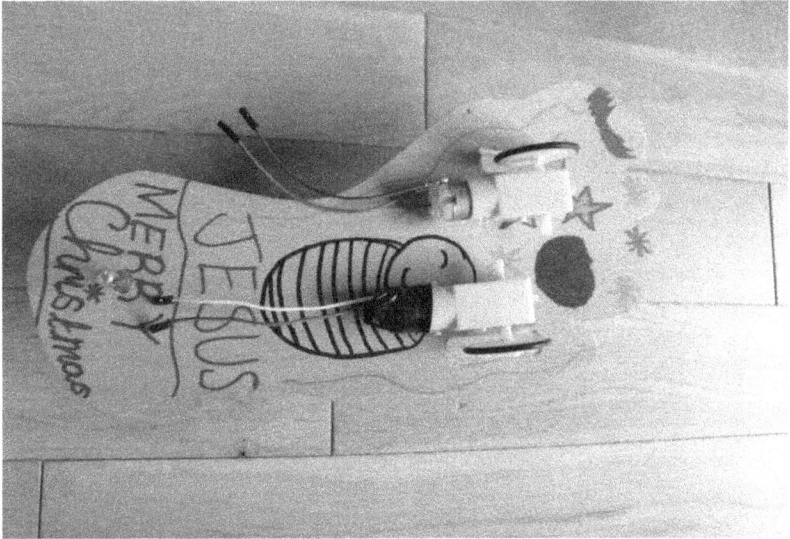

Figure 8.9: Robotics Five Toes Foot colour Baby Jesus

J. Robotics Five Toes Foot colour purpose waves

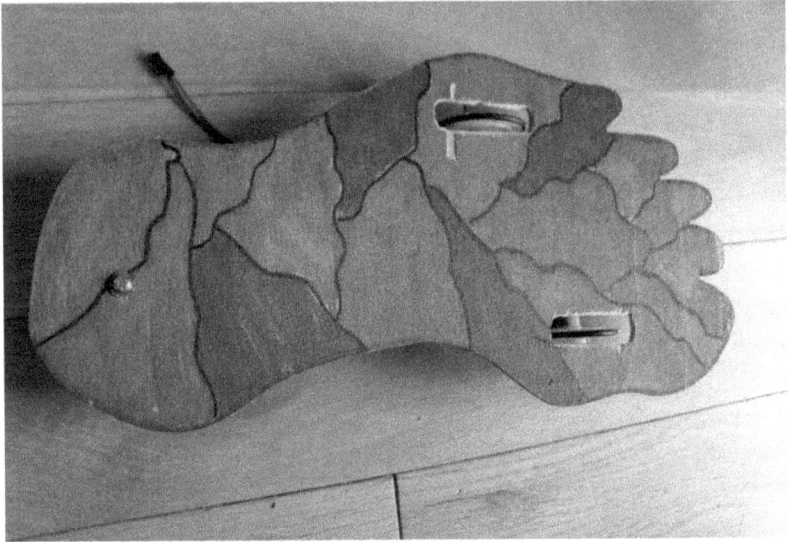

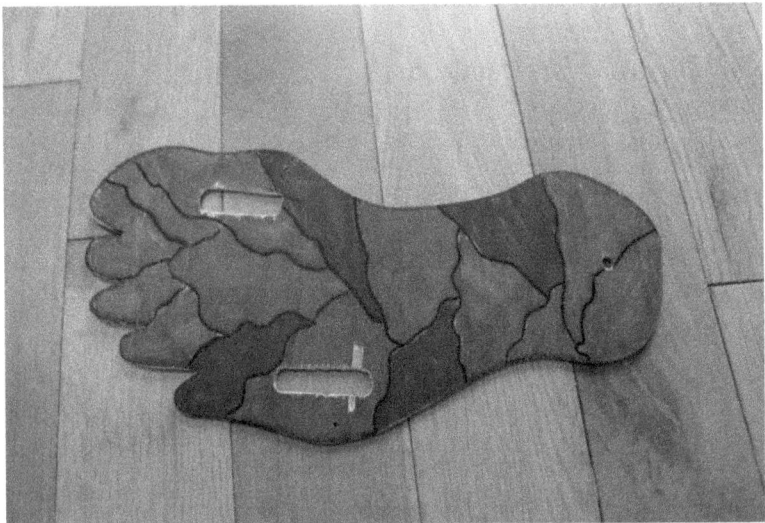

Figure 8.10: Robotics Five Toes Foot colour purpose
waves

K. Robotics Five Toes Foot colour green waves

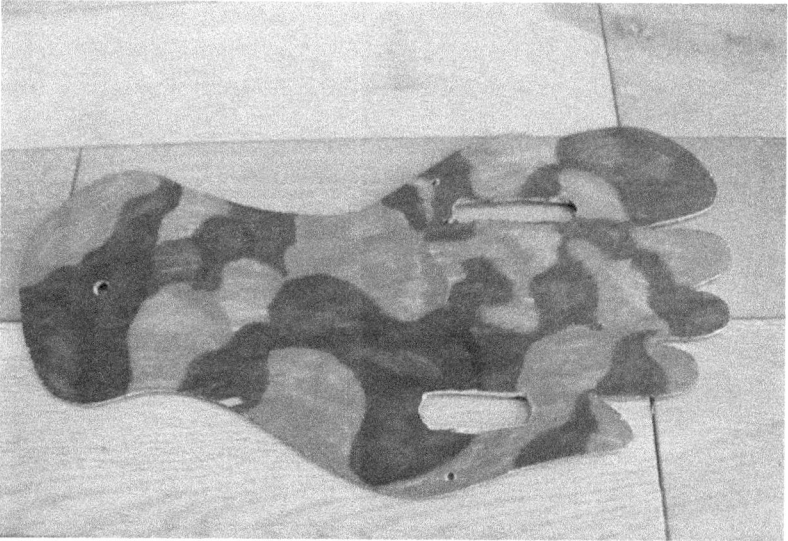

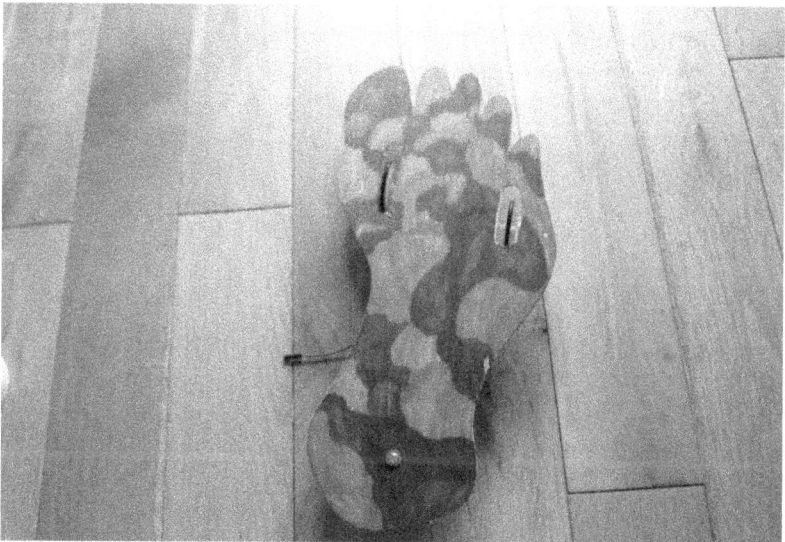

Figure 8.11: Robotics Five Toes Foot colour green waves

K. Robotics Five Toes Foot colour tracks

Figure 8.12: Robotics Five Toes Foot colour tracks

CHAPTER 9

Creative Robotics

A. Creative Robotics Model 1

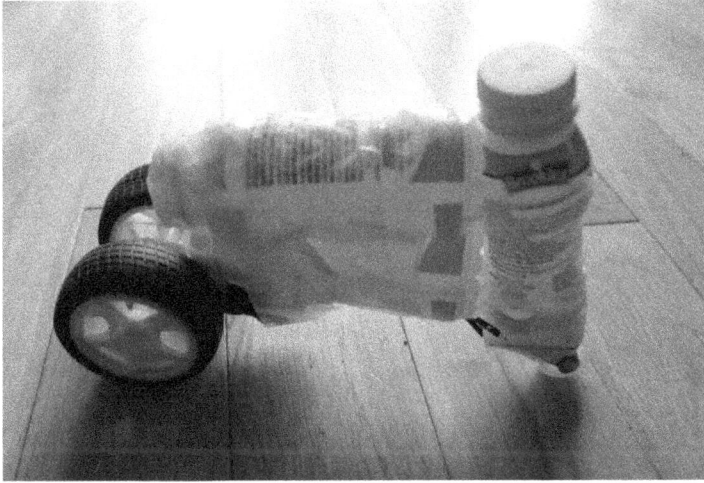

Figure 9.1: Creative Robotics Model 1

B. Creative Robotics Model 2

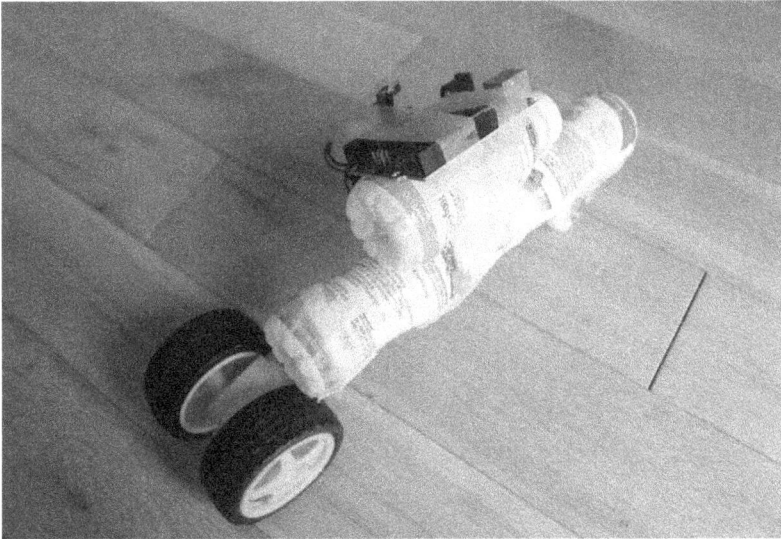

Figure 9.2: Creative Robotics Model 2

C. Creative Robotics Model 3

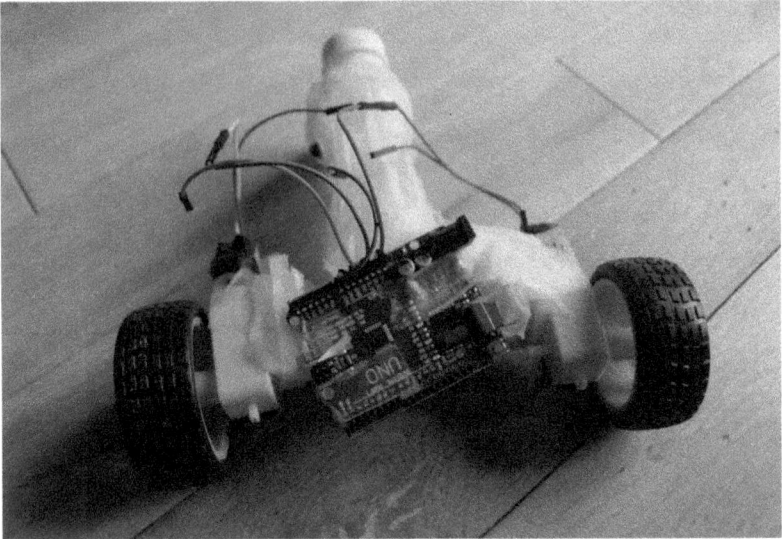

Figure 9.3: Creative Robotics Model 3

Create Educational Robotics

D. Creative Robotics Model 4

Figure 9.4: Creative Robotics Model 4

E. Creative Robotics Model 5

Figure 9.5: Creative Robotics Model 5

REFERENCES

https://www.arduino.cc/

http://www.ronsek.com/

Create Educational Robotics